Data Analytics

Thomas A. Runkler

Data Analytics

Models and Algorithms for
Intelligent Data Analysis

Third Edition

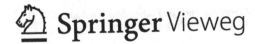 Springer Vieweg

Thomas A. Runkler
München, Germany

ISBN 978-3-658-29778-7   ISBN 978-3-658-29779-4   (eBook)
https://doi.org/10.1007/978-3-658-29779-4

This Springer Vieweg imprint is published by the registered company Springer Fachmedien Wiesbaden GmbH.
The registered company address is: Abraham-Lincoln-Str. 46, 65189 Wiesbaden, Germany

Preface

The information in the world doubles every 20 months. Important data sources are business and industrial processes, text and structured databases, images and videos, and physical and biomedical data. Data analytics allows to find relevant information, structures, and patterns, to gain new insights, to identify causes and effects, to predict future developments, or to suggest optimal decisions. We need models and algorithms to collect, preprocess, analyze, and evaluate data. This involves methods from various fields such as statistics, machine learning, pattern recognition, systems theory, operations research, or artificial intelligence. With this book, you will learn about the most important methods and algorithms for data analytics. You will be able to choose appropriate methods for specific tasks and apply these in your own data analytics projects. You will understand the basic concepts of the growing field of data analytics, which will allow you to keep pace and to actively contribute to the advancement of the field.

This text is designed for undergraduate and graduate courses on data analytics for engineering, computer science, and math students. It is also suitable for practitioners working on data analytics projects. The book is structured according to typical practical data analytics projects. Only basic mathematics is required. This material has been used for many years in courses at the Technical University of Munich, Germany, and at many other universities. Much of the content is based on the results of industrial research and development projects at Siemens.

The following list shows the history of the book versions:

- Data Analytics, third edition, 2020, English
- Data Analytics, second edition, 2016, English
- Data Mining, second edition, 2015, German
- Data Analytics, 2012, English
- Data Mining, 2010, German
- Information Mining, 2000, German

Thanks to everybody who has contributed to this material, in particular to the reviewers and students for suggesting improvements and pointing out errors and to the editorial and publisher team for their professional collaboration.

Munich, Germany   Thomas A. Runkler
April 2020

Contents

List of Symbols

$\forall x \in X$	For each x in X
$\exists x \in X$	There exists an x in X
$\Rightarrow$	If ... then ...
$\Leftrightarrow$	If and only if
$\int_a^b f \, dx$	Integral of f from $x = a$ to $x = b$
$\frac{\partial f}{\partial x}$	Partial derivative of f with respect to x
$\wedge$	Conjunction
$\vee$	Disjunction
$\cap$	Intersection
$\cup$	Union
$\neg$	Complement
$\backslash$	Set difference
$\subset, \subseteq$	Inclusion
$\cdot$	Product, inner product
$\times$	Cartesian product, vector product
$\{\}$	Empty set
$[x, y]$	Closed interval from x to y
$(x, y], [x, y)$	Half-bounded intervals from x to y
(x, y)	Open interval from x to y
$|x|$	Absolute value of x
$|X|$	Cardinality of the set X
$\|x\|$	Norm of vector x
$\lfloor x \rfloor$	Smallest integer $a \geq x$
$\lceil x \rceil$	Largest integer $a \leq x$
$\binom{n}{m}$	Vector with the components n and m, binomial coefficient
∞	Infinity
$a \ll b$	a is much less than b
$a \gg b$	a is much greater than b
$\alpha(t)$	Time variant learning rate
argmin X	Index of the minimum of X

argmax X	Index of the maximum of X
arctan x	Arctangent of x
artanh x	Inverse hyperbolic tangent of x
c_{ij}	Covariance between features i and j
$CE(U)$	Classification entropy of U
cov X	Covariance matrix of X
$d(a, b)$	Distance between a and b
eig X	Eigenvectors and eigenvalues of X
F_c	Fourier cosine transform
F_s	Fourier sine transform
$h(X)$	Hopkins index of X
$H(a, b)$	Hamming distance between a and b
$H(Z)$	Minimal hypercube or entropy of Z
$H(Z \mid a)$	Entropy of Z given a
inf X	Infimum of X
λ	Eigenvalue, Lagrange variable
$L(a, b)$	Edit distance between a and b
$\lim_{x \to y}$	Limit as x approaches y
$\log_b x$	Logarithm of x to base b
max X	Maximum of X
min X	Minimum of X
a mod b	a modulo b
$N(\mu, \sigma)$	Gaussian distribution with mean μ and standard deviation σ
NaN	Undefined (not a number)
$PC(U)$	Partition coefficient of U
$\mathbb{R}$	Set of real numbers
$\mathbb{R}^+$	Set of positive real numbers
r	Radius
s	Standard deviation
s_{ij}	Correlation between features i and j
sup X	Supremum of X
tanh x	Hyperbolic tangent of x
u_{ik}	Membership of the kth vector in the ith cluster
X	Set or matrix X
$\bar{x}$	Average of X
X^T, x^T	Transpose of the matrix X, or the vector x
x_k	k^{th} vector of X
$x^{(i)}$	i^{th} component of X
$x_k^{(i)}$	i^{th} component of the kth vector of X
x	Scalar or vector x
$x(t)$	Time signal
$x(j2\pi f)$	Spectrum

Introduction

Abstract

This book deals with models and algorithms for the analysis of data sets, for example industrial process data, business data, text and structured data, image data, and biomedical data. We define the terms data analytics, data mining, knowledge discovery, and the KDD and CRISP-DM processes. Typical data analysis projects can be divided into several phases: preparation, preprocessing, analysis, and postprocessing. The chapters of this book are structured according to the main methods of data preprocessing and data analysis: data and relations, data preprocessing, visualization, correlation, regression, forecasting, classification, and clustering.

1.1 It's All About Data

The focus of this book is the analysis of large data sets, for example:

- Industrial process data: An increasing amount of data is acquired, stored and processed in order to automate and control industrial production, manufacturing, distribution, logistics and supply chain processes. Data are used on all levels of the automation pyramid: sensors and actuators at the field level, control signals at the control level, operation and monitoring data at the execution level, schedules and indicators at the planning level. The main purpose of data analysis in industry is to optimize processes and to improve the competitive position of the company [5].
- Business data: Data of business performance are analyzed to better understand and drive business processes [6]. Important business domains to be analyzed include customers, portfolio, sales, marketing, pricing, financials, risk, and fraud. An example is shopping basket analysis that finds out which products customers purchase at the same time. This

analysis aims to improve cross selling. Another example for business data analysis is customer segmentation for tailored advertising and sales promotions.

- Text data: The analysis of numerical data has been the focus of mathematical statistics for centuries. Today, text data also serve as important information sources, for example text documents, text messages, or web documents. Text analysis helps to filter, search, extract, and structure information, with the overall goal to process and understand natural language [8]. This is often enhanced by semantic models [1], ontologies [14], and knowledge graphs [2].
- Image data: An increasing number of camera sensors ranging from smartphones, traffic surveillance, to image satellites yields large amounts of 2D and also 3D image and video data. Image and video analysis finds and recognizes objects, analyzes and classifies scenes, and relates image data with other information sources [13].
- Biomedical data: Data from laboratory experiments are used to analyze, understand and exploit biological and medical processes, for example to analyze DNA sequences, to understand and annotate genome functions, to analyze gene and protein expressions or to model regulation networks [10].

1.2   Data Analytics, Data Mining, and Knowledge Discovery

The term *data mining* dates back to the 1980s [7]. The goal of data mining is to extract knowledge from data [3]. In this context, *knowledge* is defined as *interesting* patterns that are generally valid, novel, useful, and understandable to humans. Whether or not the extracted patterns are interesting depends on the particular application and needs to be verified by application experts. Based on expert feedback the knowledge extraction process is often interactively refined. The term *data analytics* first appeared in the early 2000s [4, 15]. Data analytics is defined as the application of computer systems to the analysis of large data sets for the support of decisions. Data analytics is a very interdisciplinary field that has adopted aspects from many other scientific disciplines such as statistics, machine learning, pattern recognition, system theory, operations research, or artificial intelligence.

Typical data analysis projects can be divided into several phases. Data are assessed and selected, cleaned and filtered, visualized and analyzed, and the analysis results are finally interpreted and evaluated. The *knowledge discovery in databases* (KDD) process [3] comprises the six phases selection, preprocessing, transformation, data mining, interpretation, and evaluation. The *cross industry standard process for data mining* (CRISP-DM) [12] comprises the six phases business understanding, data understanding, data preparation, modeling, evaluation, and deployment. For simplicity we distinguish only four phases here: preparation, preprocessing, analysis, and postprocessing (Fig. 1.1). The main focus of this book is data preprocessing and data analysis. The chapters are structured according to the main methods of preprocessing and analysis: data and relations, data preprocessing, visualization, correlation, regression, forecasting, classification, and clustering.

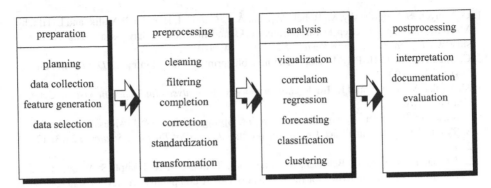

Fig. 1.1 Phases of data analysis projects

This book gives a clear and concise overview of the most important methods and algorithms of data analysis. It enables the reader to gain a complete and comprehensive understanding of data analysis, to apply data analysis methods to her or his own projects, and to contribute to the progress of the field.

A large number a software tools for data mining are available today. Popular commercial or free software tools include Alteryx, AutoML, KNIME, MATLAB, Python, R, Rapid Miner, SAS, Spark, Splunk, SPSS, TIBCO Spotfire, Tableau, QlikView, or WEKA. This book does *not* present, compare, or recommend any data mining software tools. For a comprehensive overview of current data mining software tools please refer to [9, 11, 16].

References

1. J. Davies, R. Studer, and P. Warren. *Semantic Web technologies: Trends and research in ontology–based systems.* John Wiley & Sons, 2006.
2. L. Ehrlinger and W. Wöß. Towards a definition of knowledge graphs. *International Conference on Semantic Systems, Leipzig, Germany,* 2016.
3. U. M. Fayyad, G. Piatetsky-Shapiro, P. Smyth, and R. Uthurusamy. *Advances in Knowledge Discovery and Data Mining.* AAAI Press, Menlo Park, 1996.
4. R. Kohavi, N. J. Rothleder, and E. Simoudis. Emerging trends in business analytics. *Communications of the ACM,* 45(8):345–48, 2002.
5. H. Lasi, P. Fettke, H. G. Kemper, T. Feld, and M. Hoffmann. Industry 4.0. *Business & Information Systems Engineering,* 6(4):239–242, 2014.
6. J. Liebowitz. *Big data and business analytics.* CRC press, 2013.
7. M. C. Lovell. Data mining. *Review of Economics and Statistics,* 65(1):1–11, 1983.
8. C. D. Manning and H. Schütze. *Foundations of Statistical Natural Language Processing.* MIT Press, 1999.
9. R. Mikut and M. Reischl. Data mining tools. *Wiley Interdisciplinary Reviews: Data Mining and Knowledge Discovery,* 1(5):431–443, 2011.
10. D. W. Mount. *Bioinformatics: Sequence and genome analysis.* Cold Spring Harbor Laboratory Press, New York, 2004.

11. G. Nguyen, S. Dlugolinsky, M. Bobák, V. Tran, Á. L. García, I. Heredia, P. Malík, and L. Hluchỳ. Machine learning and deep learning frameworks and libraries for large–scale data mining: A survey. *Artificial Intelligence Review*, 52(1):77–124, 2019.

12. C. Shearer. The CRISP-DM model: The new blueprint for data mining. *J Data Warehousing*, 5(4):13–22, 2000.

13. M. Sonka, V. Hlavac, and R. Boyle. *Image processing, analysis, and machine vision*. Cengage Learning, 2014.

14. S. Staab and R. Studer. *Handbook on ontologies*. Springer Science & Business Media, 2013.

15. S. Tyagi. Using data analytics for greater profits. *Journal of Business Strategy*, 24(3):12–14, 2003.

16. L. Zhang, A. Stoffel, M. Behrisch, S. Mittelstädt, T. Schreck, R. Pompl, S. Weber, H. Last, and D. Keim. Visual analytics for the big data era — A comparative review of state-of-the-art commercial systems. In *IEEE Conference on Visual Analytics Science and Technology*, pages 173–182, Seattle, 2012.

Data and Relations

2

Abstract

The popular Iris benchmark set is used to introduce the basic concepts of data analysis. Data scales (nominal, ordinal, interval, ratio) must be considered because certain mathematical operations are only appropriate for specific scales. Numerical data can be represented by sets, vectors, or matrices. Data analysis is often based on dissimilarity measures (like matrix norms, Lebesgue/Minkowski norms) or on similarity measures (like cosine, overlap, Dice, Jaccard, Tanimoto). Sequences can be analyzed using sequence relations (like Hamming or edit distance). Data can be extracted from continuous signals by sampling and quantization. The Nyquist condition allows sampling without loss of information.

2.1 The Iris Data Set

To introduce the basic concepts of data analysis we consider one of the most popular historic benchmark data sets: the *Iris* data set [1]. The Iris data set was originally created in 1935 by the American botanist Edgar Anderson who examined the geographic distribution of Iris flowers on the Gaspé peninsula in Quebec (Canada). In 1936, Sir Ronald Aylmer Fisher used Anderson's Iris data set as an example for multivariate discriminant analysis [4] (compare Chap. 8). Subsequently, the Iris data set became one of the most frequently used reference data sets in statistics and data analysis.

The Iris data set comprises measurements from 150 Iris flower samples: 50 from each of the three species Iris Setosa, Iris Virginica, and Iris Versicolor. For each of the 150 flowers, values of four numerical features chosen by Anderson were measured: the length and the width of sepal and petal leaves in centimeters. For illustration, Table 2.1 shows the complete Iris data set. Notice that several *distinct* replicates of the original Iris data set have been used and published, so in experiments with this data set the version should be

© Springer Fachmedien Wiesbaden GmbH, part of Springer Nature 2020
T. A. Runkler, *Data Analytics*, https://doi.org/10.1007/978-3-658-29779-4_2

Table 2.1 The Iris data set (from [1])

Setosa				Versicolor				Virginica			
Sepal		Petal		Sepal		Petal		Sepal		Petal	
Length	Width	Length	Width	Length	Width	Length	Width	Length	Width	Length	Width
5.1	3.5	1.4	0.2	7	3.2	4.7	1.4	6.3	3.3	6	2.5
4.9	3	1.4	0.2	6.4	3.2	4.5	1.5	5.8	2.7	5.1	1.9
4.7	3.2	1.3	0.2	6.9	3.1	4.9	1.5	7.1	3	5.9	2.1
4.6	3.1	1.5	0.2	5.5	2.3	4	1.3	6.3	2.9	5.6	1.8
5	3.6	1.4	0.2	6.5	2.8	4.6	1.5	6.5	3	5.8	2.2
5.4	3.9	1.7	0.4	5.7	2.8	4.5	1.3	7.6	3	6.6	2.1
4.6	3.4	1.4	0.3	6.3	3.3	4.7	1.6	4.9	2.5	4.5	1.7
5	3.4	1.5	0.2	4.9	2.4	3.3	1	7.3	2.9	6.3	1.8
4.4	2.9	1.4	0.2	6.6	2.9	4.6	1.3	6.7	2.5	5.8	1.8
4.9	3.1	1.5	0.1	5.2	2.7	3.9	1.4	7.2	3.6	6.1	2.5
5.4	3.7	1.5	0.2	5	2	3.5	1	6.5	3.2	5.1	2
4.8	3.4	1.6	0.2	5.9	3	4.2	1.5	6.4	2.7	5.3	1.9
4.8	3	1.4	0.1	6	2.2	4	1	6.8	3	5.5	2.1
4.3	3	1.1	0.1	6.1	2.9	4.7	1.4	5.7	2.5	5	2
5.8	4	1.2	0.2	5.6	2.9	3.6	1.3	5.8	2.8	5.1	2.4
5.7	4.4	1.5	0.4	6.7	3.1	4.4	1.4	6.4	3.2	5.3	2.3
5.4	3.9	1.3	0.4	5.6	3	4.5	1.5	6.5	3	5.5	1.8
5.1	3.5	1.4	0.3	5.8	2.7	4.1	1	7.7	3.8	6.7	2.2
5.7	3.8	1.7	0.3	6.2	2.2	4.5	1.5	7.7	2.6	6.9	2.3
5.1	3.8	1.5	0.3	5.6	2.5	3.9	1.1	6	2.2	5	1.5
5.4	3.4	1.7	0.2	5.9	3.2	4.8	1.8	6.9	3.2	5.7	2.3
5.1	3.7	1.5	0.4	6.1	2.8	4	1.3	5.6	2.8	4.9	2
4.6	3.6	1	0.2	6.3	2.5	4.9	1.5	7.7	2.8	6.7	2
5.1	3.3	1.7	0.5	6.1	2.8	4.7	1.2	6.3	2.7	4.9	1.8
4.8	3.4	1.9	0.2	6.4	2.9	4.3	1.3	6.7	3.3	5.7	2.1
5	3	1.6	0.2	6.6	3	4.4	1.4	7.2	3.2	6	1.8
5	3.4	1.6	0.4	6.8	2.8	4.8	1.4	6.2	2.8	4.8	1.8
5.2	3.5	1.5	0.2	6.7	3	5	1.7	6.1	3	4.9	1.8
5.2	3.4	1.4	0.2	6	2.9	4.5	1.5	6.4	2.8	5.6	2.1
4.7	3.2	1.6	0.2	5.7	2.6	3.5	1	7.2	3	5.8	1.6
4.8	3.1	1.6	0.2	5.5	2.4	3.8	1.1	7.4	2.8	6.1	1.9
5.4	3.4	1.5	0.4	5.5	2.4	3.7	1	7.9	3.8	6.4	2
5.2	4.1	1.5	0.1	5.8	2.7	3.9	1.2	6.4	2.8	5.6	2.2
5.5	4.2	1.4	0.2	6	2.7	5.1	1.6	6.3	2.8	5.1	1.5
4.9	3.1	1.5	0.2	5.4	3	4.5	1.5	6.1	2.6	5.6	1.4
5	3.2	1.2	0.2	6	3.4	4.5	1.6	7.7	3	6.1	2.3

(continued)

Table 2.1 (continued)

Setosa				Versicolor				Virginica			
Sepal		Petal		Sepal		Petal		Sepal		Petal	
Length	Width	Length	Width	Length	Width	Length	Width	Length	Width	Length	Width
5.5	3.5	1.3	0.2	6.7	3.1	4.7	1.5	6.3	3.4	5.6	2.4
4.9	3.6	1.4	0.1	6.3	2.3	4.4	1.3	6.4	3.1	5.5	1.8
4.4	3	1.3	0.2	5.6	3	4.1	1.3	6	3	4.8	1.8
5.1	3.4	1.5	0.2	5.5	2.5	4	1.3	6.9	3.1	5.4	2.1
5	3.5	1.3	0.3	5.5	2.6	4.4	1.2	6.7	3.1	5.6	2.4
4.5	2.3	1.3	0.3	6.1	3	4.6	1.4	6.9	3.1	5.1	2.3
4.4	3.2	1.3	0.2	5.8	2.6	4	1.2	5.8	2.7	5.1	1.9
5	3.5	1.6	0.6	5	2.3	3.3	1	6.8	3.2	5.9	2.3
5.1	3.8	1.9	0.4	5.6	2.7	4.2	1.3	6.7	3.3	5.7	2.5
4.8	3	1.4	0.3	5.7	3	4.2	1.2	6.7	3	5.2	2.3
5.1	3.8	1.6	0.2	5.7	2.9	4.2	1.3	6.3	2.5	5	1.9
4.6	3.2	1.4	0.2	6.2	2.9	4.3	1.3	6.5	3	5.2	2
5.3	3.7	1.5	0.2	5.1	2.5	3	1.1	6.2	3.4	5.4	2.3
5	3.3	1.4	0.2	5.7	2.8	4.1	1.3	5.9	3	5.1	1.8

carefully checked [2]. The Iris data set as well as many other popular data sets are available through public data bases, for example the machine learning data base at the University of California at Irvine (UCI).

In data analysis we call each of the 150 Iris flowers an *object*, each of the three species a *class*, and each of the four dimensions a *feature*. Here is a list of typical questions that we try to answer by data analysis:

- Which of the data might contain errors or false class assignments?
- What is the error caused by rounding the data off to one decimal place?
- What is the correlation between petal length and petal width?
- Which pair of dimensions is correlated most?
- None of the flowers in the data set has a sepal width of 1.8 cm. Which sepal length would we expect for a flower that did have 1.8 cm as its sepal width?
- Which species would an Iris with a sepal width of 1.8 cm belong to?
- Do the three species contain sub-species that can be identified from the data?

In this book you will find numerous methods and algorithms to answer these and other data analysis questions. For a better understanding of these data analysis methods and algorithms we first define and examine the fundamental properties of data and their relations.

2.2   Data Scales

Numerical measurements may have different semantic meanings, even if they are represented by the same numerical data. Depending on the semantic meaning different types of mathematical operations are appropriate. For the semantic meaning of numerical measurement Stevens [7] suggested the four different *scales* that are shown in Table 2.2. For nominal scaled data (bottom row) only tests for equality or inequality are valid. Examples for nominal features are names of persons or codes of objects. Data of a nominal feature can be represented by the *mode* which is defined as the value that occurs most frequently. For ordinal scaled data (third row) the operations "greater than" and "less than" are valid. For each scale level the operations and statistics of the lower scale levels are also valid, so for the ordinal scale we have equality, inequality, and the combinations "greater than or equal" ($\geq$) and "less than or equal" ($\leq$). The relation "less than or equal" ($\leq$) defines a *total order*, such that for any x, y, z we have $(x \leq y) \wedge (y \leq x) \Rightarrow (x = y)$ (antisymmetry), $(x \leq y) \wedge (y \leq z) \Rightarrow (x \leq z)$ (transitivity), and $(x \leq y) \vee (y \leq x)$ (totality). Examples for ordinal features are school grades. Data of an ordinal feature can be represented by the *median* which is defined as the value for which (almost) as many smaller as larger values exist. The *mean* is not valid for ordinal features, so for example it does not make sense to say that the average school grade is C. For interval scaled data (second row) addition and subtraction are valid. Interval scaled features have arbitrary zero points. Examples are years in the Anno Domini dating system or temperatures in degrees Celsius (centigrade) or Fahrenheit, so for example it does not make sense to say that 40 °C is twice as much as 20 °C. Data of an interval scaled feature, for example, a set of values $X = \{x_1, \ldots, x_n\}$, can be represented by the (arithmetic) *mean*

$$\bar{x} = \frac{1}{n} \sum_{k=1}^{n} x_k \tag{2.1}$$

For ratio scaled data (top row) multiplication and division are valid. Examples for ratio scaled features are time differences like ages or temperatures on the Kelvin scale. Data of an interval scaled feature can be represented by the *generalized mean*

$$m_\alpha(X) = \sqrt[\alpha]{\frac{1}{n} \sum_{k=1}^{n} x_k^\alpha} \tag{2.2}$$

Table 2.2 Scales for numerical measurements

Scale	Operations		Example	Statistics
Ratio	·	/	21 years, 273K	Generalized mean
Interval	+	−	2015 A.D., 20 °C	Mean
Ordinal	>	<	A, B, C, D, F	Median
Nominal	=	≠	Alice, Bob, Carol	Mode

Table 2.3 Computation of the median petal width of the Iris data set

Value	Count	Accumulated count	Value	Count	Accumulated count
0.1	5	5	2.5	3	3
0.2	29	34	2.4	3	6
0.3	7	41	2.3	8	14
0.4	7	48	2.2	3	17
0.5	1	49	2.1	6	23
0.6	1	50	2	6	29
0.7	0	50	1.9	5	34
0.8	0	50	1.8	12	46
0.9	0	50	1.7	2	48
1	7	57	1.6	4	52
1.1	3	60	1.5	12	64
1.2	5	65	1.4	8	72
1.3	10(13)	75	1.3	3(13)	75

with the parameter $\alpha \in \mathbb{R} \setminus \{0\}$, which includes the special cases minimum ($\alpha \to -\infty$), harmonic mean ($\alpha = -1$), geometric mean ($\alpha \to 0$), arithmetic mean ($\alpha = 1$), quadratic mean ($\alpha = 2$), and maximum ($\alpha \to \infty$).

The features of the Iris data set are on ratio scale. For example, we can approximately estimate the sepal surface area by multiplying the sepal length and the sepal width. Hence, we can compute the mode, median, mean and generalized mean of each of the features of the Iris data set. Table 2.3 illustrates the computation of the mode and the median of the petal width (fourth feature). The Iris data set contains petal widths between 0.1 and 2.5 cm. The most frequent value of the petal width is 0.2 cm, which occurs 29 times, so the mode is 0.2 cm. To compute the median we can accumulate the numbers of occurrences (counts) of the values for 0.1 cm, 0.2 cm, and so on, until we reach half of the number of objects (75). This algorithm yields a median petal width of 1.3 cm. The accumulation of the counts can also be done in reverse order (right view of Table 2.3). The complexity of this algorithm is $O(n \log n)$. Notice, however, and this is surprising even to many scientists, that the median can be efficiently computed in linear time using selection algorithms [3].

Many methods presented in this book use addition and subtraction, or even multiplication and division of feature values, and are therefore suitable only for interval or ratio scaled data, respectively. To analyze nominal and ordinal data we may define relations between pairs of such data, that can be analyzed using specific relational methods.

2.3   Set and Matrix Representations

We denote numerical feature data as the set

$$X = \{x_1, \ldots, x_n\} \subset \mathbb{R}^p \tag{2.3}$$

with n elements, where each element is a p-dimensional real-valued feature vector, where n and p are positive integers. For $p = 1$ we call X a *scalar* data set. As an alternative to the set representation, numerical feature data are also often represented as a matrix

$$X = \begin{pmatrix} x_1^{(1)} & \cdots & x_1^{(p)} \\ \vdots & \ddots & \vdots \\ x_n^{(1)} & \cdots & x_n^{(p)} \end{pmatrix} \tag{2.4}$$

so the vectors $x_1, \ldots, x_n$ are *row* vectors. Although mathematically a bit sloppy, data sets and data matrices are commonly used as equivalent data representations. Figure 2.1 illustrates the common terms and notations of a data matrix. Each row of the data matrix corresponds to an element of the data set. It is called *feature vector* or *data point* x_k, $k = 1, \ldots, n$. Each column of the data matrix corresponds to one component of all elements of the data set. It is called i^{th} *feature* or i^{th} *component* $x^{(i)}$, $i = 1, \ldots, p$. In this book we distinguish rows and columns by using subscripts for rows and bracketed superscripts for columns. Alternative notations in the literature are $x(k, .)$ and $x(., i)$, for example. A single matrix element is a component of an element of the data set. It is called *datum* or *value* $x_k^{(i)}$, $k = 1, \ldots, n$, $i = 1, \ldots, p$.

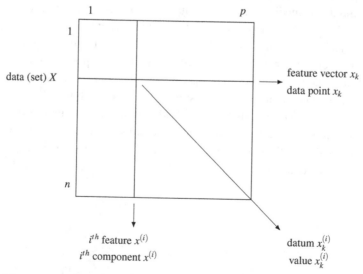

Fig. 2.1 Matrix representation of a data set

The Iris data set can be written as a data matrix with 150 rows and 4 columns, where each row represents one object (flower) and each column represents one feature (dimension). The Iris data matrix can be obtained by vertical concatenation of the three portions shown in Table 2.1. The class information (Setosa, Versicolor, Virginica) can be interpreted as a fifth feature, on nominal scale.

2.4 Relations

Consider a set of (abstract) elements, without referring to numerical feature vectors.

$$O = \{o_1, \ldots, o_n\} \tag{2.5}$$

Sometimes no feature vector representation is available for the objects o_k, $k = 1, \ldots, n$, so conventional feature-based data analysis methods are not applicable. Instead, the relation of all pairs of objects can often be quantified and written as a square matrix

$$R = \begin{pmatrix} r_{11} & \cdots & r_{1n} \\ \vdots & \ddots & \vdots \\ r_{n1} & \cdots & r_{nn} \end{pmatrix} \in \mathbb{R}^{n \times n} \tag{2.6}$$

Each relation value r_{ij}, $i, j = 1, \ldots, n$, may refer to a degree of similarity, dissimilarity, compatibility, incompatibility, proximity or distance between the pair of objects o_i and o_j. R may be symmetric, so $r_{ij} = r_{ji}$ for all $i, j = 1, \ldots, n$. R may be manually defined or computed from features. If numerical features X are available, then R may be computed from X using an appropriate function $f : \mathbb{R}^p \times \mathbb{R}^p \to \mathbb{R}$. For example, a relational matrix for Iris may be manually defined by a botanist who optically compares and then numerically scores some relationship between pairs of flowers, or R may be computed from sepal and petal lengths and widths. Two important classes of relations, dissimilarities and similarities, are presented in the following two sections.

2.5 Dissimilarity Measures

A function d is called *dissimilarity* or *distance measure* if for all $x, y \in \mathbb{R}^p$

$$d(x, y) = d(y, x) \tag{2.7}$$

$$d(x, y) = 0 \quad \Leftrightarrow \quad x = y \tag{2.8}$$

$$d(x, z) \leq d(x, y) + d(y, z) \tag{2.9}$$

From these axioms follows

$$d(x, y) \geq 0 \qquad (2.10)$$

A class of dissimilarity measures is defined using a *norm* $\|.\|$ of $x - y$, so

$$d(x, y) = \|x - y\| \qquad (2.11)$$

A function $\|.\| : \mathbb{R}^p \to \mathbb{R}^+$ is a norm if and only if

$$\|x\| = 0 \Leftrightarrow x = (0, \ldots, 0) \qquad (2.12)$$

$$\|a \cdot x\| = |a| \cdot \|x\| \quad \forall a \in \mathbb{R}, x \in \mathbb{R}^p \qquad (2.13)$$

$$\|x + y\| \leq \|x\| + \|y\| \quad \forall x, y \in \mathbb{R}^p \qquad (2.14)$$

For example, the frequently used so-called *hyperbolic norm*

$$\|x\|_h = \prod_{i=1}^{p} x^{(i)} \qquad (2.15)$$

is not a norm according to the previous definition, since condition (2.12) is violated by $x = (0, 1) \neq (0, 0)$ with $\|x\|_h = \|(0, 1)\|_h = 0$, or condition (2.13) is violated by $x = (1, 1)$ and $a = 2$, where $\|a \cdot x\|_h = \|2 \cdot (1, 1)\|_h = \|(2, 2)\|_h = 4 \neq |a| \cdot \|x\|_h = |2| \cdot \|(1, 1)\|_h = 2$.

Frequently used classes of norms are *matrix norms* and *Lebesgue* or *Minkowski norms*. The matrix norm is defined as

$$\|x\|_A = \sqrt{xAx^T} \qquad (2.16)$$

with a matrix $A \in \mathbb{R}^{n \times n}$. Important special cases of the matrix norm are the *Euclidean norm*

$$A = \begin{pmatrix} 1 & 0 & \cdots & 0 \\ 0 & 1 & \cdots & 0 \\ \vdots & \vdots & \ddots & \vdots \\ 0 & 0 & \cdots & 1 \end{pmatrix} \qquad (2.17)$$

the *Frobenius* or *Hilbert-Schmidt norm*

$$A = \begin{pmatrix} 1 & 1 & \cdots & 1 \\ 1 & 1 & \cdots & 1 \\ \vdots & \vdots & \ddots & \vdots \\ 1 & 1 & \cdots & 1 \end{pmatrix} \tag{2.18}$$

the *diagonal norm* with individual weights for each feature

$$A = \begin{pmatrix} d_1 & 0 & \cdots & 0 \\ 0 & d_2 & \cdots & 0 \\ \vdots & \vdots & \ddots & \vdots \\ 0 & 0 & \cdots & d_p \end{pmatrix} \tag{2.19}$$

and the *Mahalanobis norm*

$$A = \text{cov}^{-1} X = \left(\frac{1}{n-1} \sum_{k=1}^{n} (x_k - \bar{x})^T (x_k - \bar{x}) \right)^{-1} \tag{2.20}$$

The Mahalanobis norm uses the inverse of the covariance matrix of the data set X (see Chap. 5). It adapts the weighting of the individual features based on the observed statistics and also accounts for correlations between pairs of features. Pairs of data at opposite ends of the observed data distribution have the same Mahalanobis distance.

The Lebesgue or Minkowski norm is defined as

$$\|x\|_\alpha = \sqrt[\alpha]{\sum_{j=1}^{p} |x^{(j)}|^\alpha} \tag{2.21}$$

with $\alpha \in \mathbb{R} \setminus \{0\}$, which is equal to the generalized mean at (2.2) except for a constant factor $\sqrt[\alpha]{n}$. Important special cases of the Lebesgue or Minkowski norm are the infimum norm $(\alpha \to -\infty)$

$$\|x\|_{-\infty} = \min_{j=1,\dots,p} x^{(j)} \tag{2.22}$$

the *Manhattan* or *city block distance* $(\alpha = 1)$

$$\|x\|_1 = \sum_{j=1}^{p} |x^{(j)}| \tag{2.23}$$

the *Euclidean norm* ($\alpha = 2$), which is the unique point in the intersection of the matrix norm and Lebesgue/Minkowski norm families,

$$\|x\|_2 = \sqrt{\sum_{j=1}^{p} \left(x^{(j)}\right)^2} \tag{2.24}$$

and the supremum norm ($\alpha \to \infty$)

$$\|x\|_\infty = \max_{j=1,\dots,p} x^{(j)} \tag{2.25}$$

Another frequently used dissimilarity measure is the *Hamming distance* [5] defined as

$$d_H(x, y) = \sum_{i=1}^{p} \rho(x^{(i)}, y^{(i)}) \tag{2.26}$$

with the discrete metric

$$\rho(x, y) = \begin{cases} 0 & \text{if } x = y \\ 1 & \text{otherwise} \end{cases} \tag{2.27}$$

So the Hamming distance yields the number of feature values that do not match. For binary features, the Hamming distance is equal to the Manhattan or city block distance (2.23), $d_H(x, y) = \|x - y\|_1$. Notice, however, that the Hamming distance is *not* associated with a norm because condition (2.13) does not hold. Variants of the Hamming distance use modified functions ρ to specify similarities between individual features. For example, if the features are (nominal scale) web pages, then ρ might be lower for pairs of pages with similar content and higher for pairs of pages with rather dissimilar content.

2.6   Similarity Measures

A function s is called *similarity* or *proximity measure* if for all $x, y \in \mathbb{R}^p$

$$s(x, y) = s(y, x) \tag{2.28}$$

$$s(x, y) \leq s(x, x) \tag{2.29}$$

$$s(x, y) \geq 0 \tag{2.30}$$

The function s is called *normalized* similarity measure if additionally

$$s(x, x) = 1 \qquad (2.31)$$

Any dissimilarity measure d can be used to define a corresponding similarity measure s and vice versa, for example using a monotonically decreasing positive function f with $f(0) = 1$ such as

$$s(x, y) = \frac{1}{1 + d(x, y)} \qquad (2.32)$$

However, the examples presented in the previous section are mostly used in their dissimilarity version, and the examples presented in this section are mostly used in their similarity version. Let us first consider similarities between binary feature vectors. A pair of binary feature vectors can be considered similar if many *ones* coincide. This conjunction can be represented by the product, so the scalar product of the feature vectors is a reasonable candidate for a similarity measure. Also for non-negative real-valued features $x, y \in (\mathbb{R}^+)^p$ similarity measures can be defined based on scalar products that may be normalized in different ways:

- cosine

$$s(x, y) = \frac{\sum\limits_{i=1}^{p} x^{(i)} y^{(i)}}{\sqrt{\sum\limits_{i=1}^{p} \left(x^{(i)}\right)^2 \sum\limits_{i=1}^{p} \left(y^{(i)}\right)^2}} \qquad (2.33)$$

- overlap

$$s(x, y) = \frac{\sum\limits_{i=1}^{p} x^{(i)} y^{(i)}}{\min\left(\sum\limits_{i=1}^{p} \left(x^{(i)}\right)^2, \sum\limits_{i=1}^{p} \left(y^{(i)}\right)^2\right)} \qquad (2.34)$$

- Dice

$$s(x, y) = \frac{2 \sum\limits_{i=1}^{p} x^{(i)} y^{(i)}}{\sum\limits_{i=1}^{p} \left(x^{(i)}\right)^2 + \sum\limits_{i=1}^{p} \left(y^{(i)}\right)^2} \qquad (2.35)$$

- Jaccard (or sometimes called Tanimoto)

$$s(x, y) = \frac{\sum\limits_{i=1}^{p} x^{(i)} y^{(i)}}{\sum\limits_{i=1}^{p} \left(x^{(i)}\right)^2 + \sum\limits_{i=1}^{p} \left(y^{(i)}\right)^2 - \sum\limits_{i=1}^{p} x^{(i)} y^{(i)}} \tag{2.36}$$

These expressions are undefined for zero feature vectors because the denominators are zero then, so the similarity has to be explicitly defined for this case, for example as zero. Notice that the cosine similarity is invariant against (positive) scaling of the feature vectors and therefore considers the relative distribution of the features,

$$s(c \cdot x, y) = s(x, y) \tag{2.37}$$

$$s(x, c \cdot y) = s(x, y) \tag{2.38}$$

for all $x, y \in \mathbb{R}^p$ and $c > 0$. Consider for example two cake recipes, one with 3 eggs, 1 1/2 cups sugar, 1 1/2 cups flour, 1/2 cup butter, and the other one with 6 eggs, 3 cups sugar, 3 cups flour, and 1 cup butter. Obviously, both recipes yield the same cake, but the second one yields twice as much cake as the first one. Following the intuitive expectation the cosine similarity between the two recipes is equal to one.

In this section we discussed functions to quantify the similarity between feature vectors (rows of the data matrix). In Chap. 5 we will discuss functions to quantify the similarity between *features* (*columns* of the data matrix), the so-called *correlation*. If the data matrix is transposed (the rows and columns are exchanged), then the correlation can also be used as an alternative way to quantify the similarity between feature vectors.

2.7   Sequence Relations

The dissimilarity and similarity measures presented in the previous two sections apply to feature vectors whose elements refer to different features. In this section we consider *sequences* of feature values or feature vectors, for example sequences of daily temperature values, text documents (sequences of alphanumerical characters), or sequences of visited web pages. Formally, such sequences could be viewed as feature vectors, but it is more appropriate to explicitly consider the sequential character, the fact that each element of the sequence refers to the same feature, and to be able to compare sequences of different lengths.

We use a function ρ to compare individual pairs of sequence elements. An example is the binary inequality function (2.27) that is used in the Hamming distance (2.26). The Hamming distance may be used as a sequence relation, but only for sequences of the same length. To compute the relation between sequences of *different* lengths, neutral elements like zeros or space characters might be appended to the shorter sequence. Depending on the alignment of the subsequences lower Hamming distances might be achieved by prepending or even optimally inserting neutral elements. This is the idea of the *Levenshtein* or *edit distance* [6] that determines the minimum number of edit operations (insert, delete, or change a sequence element) necessary to transform one sequence into the other. We denote $L_{ij}(x, y)$ as the edit distance between the first i elements of x and the first j elements of y, and recursively define the edit distance as

$$
L_{ij} = \begin{cases} i & j = 0 \\ j & i = 0 \\ \min\{L_{i-1, j} + 1, \ L_{i, j-1} + 1, \ L_{i-1, j-1} + \rho(x^{(i)}, y^{(j)})\} & \text{otherwise} \end{cases} \tag{2.39}
$$

The first two cases consider empty sequences and terminate the recursion. In the third case, the three arguments of the minimum operator correspond to the three edit operations insert, delete, and change. To compute L_{ij} we have to compute all L_{st}, $s = 1, \ldots, i$, $t = 1, \ldots, j$. A direct implementation of the recursive definition of the edit distance is inefficient because it computes many values of L_{st} multiple times. Figure 2.2 shows an efficient iterative implementation of the edit distance that computes each L_{st} only once. This algorithm first initializes the distances of the empty sequences $L_{s0} = s, s = 1, \ldots, i$, and $L_{0t} = t, t = 1, \ldots, j$, and then iteratively computes new distance values L_{st} from the already computed distances $L_{s-1\,t}$, $L_{s\,t-1}$, and $L_{s-1\,t-1}$. The algorithm in Fig. 2.2 computes L column by column. Instead, L can also be computed row by row or in a diagonal scheme.

Figure 2.3 shows the edit distance matrix L for the alphanumerical character sequences CAESAR and CLEOPATRA. Each matrix element is computed as the minimum of its top neighbor plus one, left neighbor plus one, and top left diagonal neighbor plus the corresponding character distance. The resulting edit distance between both sequences in the bottom right value, in this case 5. This means, that we need at least five edit operations to convert the sequence CAESAR into the sequence CLEOPATRA and vice versa. Figure 2.4 shows five such edit operations: change character 2, change character 4, insert character 5, insert character 7, and insert character 9.

Fig. 2.2 An iterative algorithm to compute the edit distance L_{ij}

```
for s = 1,...,i
    L_{s0} = s
for t = 1,...,j
    L_{0t} = t

for s = 1,...,i
    for t = 1,...,j
        L_{st} = min{L_{s-1t}  +1,
                     L_{st-1}  +1,
                     L_{s-1t-1}+ρ(x^{(s)},y^{(t)})}
```

		C	L	E	O	P	A	T	R	A
	0	1	2	3	4	5	6	7	8	9
C	1	0	1	2	3	4	5	6	7	8
A	2	1	1	2	3	4	4	5	6	7
E	3	2	2	1	2	3	4	5	6	7
S	4	3	3	2	2	3	4	5	6	7
A	5	4	4	3	3	3	3	4	5	6
R	6	5	5	4	4	4	4	4	4	5

Fig. 2.3 Edit distance matrix for the sequences CAESAR and CLEOPATRA

C	A	E	S		A		R		
0	1	0	1	1	0	1	0	1
C	L	E	O	P	A	T	R	A

$\Rightarrow 5$

Fig. 2.4 Edit operations to convert the sequence CAESAR into the sequence CLEOPATRA

2.8 Sampling and Quantization

In the previous section we have considered finite discrete sequences of features or feature vectors. In many cases such a sequence is obtained by sampling a continuous signal $x(t)$ with a fixed sampling period T, for example by measuring the room temperature every 10 min, so we obtain the time series

$$x_k = x(k \cdot T), \quad k = 1, \ldots, n \tag{2.40}$$

Figure 2.5 shows an example of such a signal and the corresponding sampled time series (vertical bars starting at zero). The time series contains only individual samples of the continuous signal, but it does not contain any information about the infinitely many points of the signal between any pair of adjacent samples, so it may cover only a part of the information contained in the signal. Increasing the sampling period will decrease

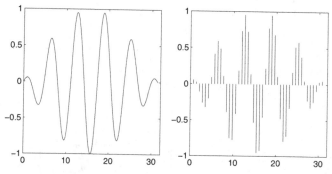

Fig. 2.5 Continuous signal and sampled time series

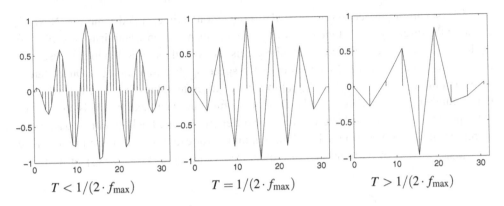

Fig. 2.6 Sampled time series for different sampling periods and reconstructed signals

the information of the time series until the time series will not provide any valuable information about the signal any more. Decreasing the sampling period may lead to a more accurate match between the time series and the signal, but it will also increase the required amount of memory to store the samples. To find a good compromise between these extremes we want to find the largest possible sampling period for which the time series contains all relevant information about the time series.

Any finite continuous signal can be represented as a sum of periodic signals with different frequencies (compare Chap. 4). A signal $x(t)$ is called *band limited* if the maximum frequency f_{max} of these periodic signals is finite, so the *Fourier spectrum* is $|x(j2\pi f)| = 0$ for $f > f_{max}$. If this signal is sampled with a sampling period less than $T_s = 1/(2 \cdot f_{max})$, or equivalently a sampling frequency larger than $f_s = 2 \cdot f_{max}$, then the original signal can be completely reconstructed from the (infinite) time series. This is called *Shannon's sampling theorem*. The condition $T \leq T_s$ (or $f \geq f_s$) is called the *Nyquist condition*. Figure 2.6 shows three time series obtained from the original signal at the left view of Fig. 2.5 using three different sampling periods. The left time series is sampled with $T < 1/(2 \cdot f_{max})$, so the Nyquist condition holds, and the piecewise linear reconstruction of the signal matches well the original signal. The middle time series is sampled with $T = 1/(2 \cdot f_{max})$, so it is at the edge of the Nyquist condition, and the reconstruction still matches the original signal. Notice however, that the match would be worse if the sampling times were delayed (corresponding to a so-called phase shift). For a delay of $T/2$, all sample values are zero in this example, yielding zero information about the original signal. The sampling period should therefore be chosen well below the Nyquist condition. Finally, the right time series is sampled with $T > 1/(2 \cdot f_{max})$, so the Nyquist condition does not hold, and much of the information of the original signal is lost.

In practical data analysis projects often only sampled data and not the original signals are available. Given only the sampled data, it is not possible to find out whether or not sampling was done according to the Nyquist condition. Therefore, it is often useful to discuss this issue with the data providers.

The first part of this section considered discretization in time, called sampling. Next we consider the discretization of the data values, called quantization. Quantization applies to analog values that are digitized with finite precision as well as digital values whose precision is reduced in order to save memory or to speed up data transmission. Quantization maps a continuous interval $[x_{min}, x_{max}]$ to a set of discrete values $\{x_1, \ldots, x_q\}$, where q is the number of quantization levels. Do not confuse the quantization levels $x_1, \ldots, x_q$ with the time series or the feature values for different objects. Each quantized value can be represented as a $b = \lceil \log_2 q \rceil$ bit binary number, for example. Each continuous value $x \in [x_{min}, x_{max}]$ can be translated to a quantized value $x_k \in \{x_1, \ldots, x_q\}$ or an index $k \in \{1, \ldots, q\}$ by rounding.

$$\frac{\frac{x_{k-1}+x_k}{2} - x_1}{x_q - x_1} \leq \frac{x - x_{min}}{x_{max} - x_{min}} < \frac{\frac{x_k+x_{k+1}}{2} - x_1}{x_q - x_1} \tag{2.41}$$

The quantization process causes a quantization error. The left view of Fig. 2.7 shows the quantization of the signal from the left view of Fig. 2.5 for $q = 11$ equidistant quantization levels $\{-1, -0.8, \ldots, 1\}$. The quantization levels appear as horizontal lines (stairs) in the curve. The right view of Fig. 2.7 shows the quantization error $e(t) = x(t) - x_q(t)$. In this case the quantization levels are equidistant, $x_i = x_1 + (i - 1) \cdot \Delta x$, $i = 1, \ldots, q$, $\Delta x = (x_q - x_1)/(q - 1)$, so the quantization error is in the interval $e(t) \in [-\Delta x/2, \Delta x/2]$. Therefore, quantization causes an additive error with the maximum absolute value $|e| \leq \Delta x/2$. To keep the quantization error low, the quantization levels should be close, i.e. Δx should be small or q should be large.

A binary number with b bits can represent integers between $x_1 = 0$ and $x_q = 2^b - 1$. The relative quantization error can then be estimated as $|e/(x_q - x_1)| \leq 100\%/2/(2^b - 1) \approx 100\%/2^{b+1}$. For typical values of $b \geq 8$ this quantization error can often be practically ignored, if the limits x_{min} and x_{max} are chosen appropriately. If the range $x_{max} - x_{min}$ is much higher than the actual variation of the data, then the quantized data might appear constant or possibly exhibit sudden jumps caused by the borders of the quantization levels. For example, if the room temperature in degrees Celsius is represented with 8 bits and

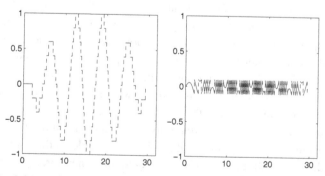

Fig. 2.7 Quantized data and quantization error

$[x_{min}, x_{max}] = [0\,°C, 1000\,°C]$, then temperatures between about $13.72\,°C$ and $17.65\,°C$ correspond to the quantization level 4, and temperatures between about $17.65\,°C$ and $21.57\,°C$ correspond to the quantization level 5. Normal room temperature will therefore yield only the values 4 or 5, and the only information contained in the quantized data is whether or not the temperature threshold of $17.65\,°C$ is exceeded.

Problems

2.1 Given a numerical data vector, for which scales can you apply the following operations?

(a) bitwise exclusive or
(b) subtraction
(c) subtraction followed by test for zero
(d) sorting in descending order
(e) computing a histogram
(f) computing the discrete Fourier cosine transform

2.2 For two-dimensional data, find the unit circle, i.e. all points with a distance of one from the origin, for

(a) Euclidean distance
(b) city block distance
(c) Hamming distance
(d) supremum norm
(e) matrix norm with $A = \begin{pmatrix} 0.1 & 0 \\ 0 & 1 \end{pmatrix}$
(f) matrix norm with $A = \begin{pmatrix} 1 & -2 \\ 0 & 1 \end{pmatrix}$

2.3 Consider sequences of the length three using the symbols $\oplus$ and $\ominus$, for example $\oplus \ominus \ominus$.

(a) Find a pair of such sequences whose Hamming distance is equal to their edit distance.
(b) Find a pair of such sequences whose Hamming distance is different from their edit distance.
(c) Find all possible sets of arbitrary number of such sequences whose Hamming distance matrices are equal to their edit distance matrices.

2.4 Consider cosine similarity for two-dimensional non-negative feature data.

(a) Find all points that have a cosine similarity of 1 with $(1, 1)$.
(b) Find all points that have a cosine similarity of $0.5\sqrt{2}$ with $(1, 1)$.
(c) Find all points that have a cosine similarity of $0.3\sqrt{10}$ with $(1, 1)$.
(d) How do you interpret these results?

References

1. E. Anderson. The Irises of the Gaspe Peninsula. *Bull. of the American Iris Society*, 59:2–5, 1935.
2. J. C. Bezdek, J. M. Keller, R. Krishnapuram, L. I. Kuncheva, and N. R. Pal. Will the real Iris data please stand up? *IEEE Transactions on Fuzzy Systems*, 7(3):368–369, 1999.
3. M. Blum, R. W. Floyd, V. Pratt, R. Rivest, and R. Tarjan. Time bounds for selection. *Journal of Computer and System Sciences*, 7:488–461, 1973.
4. R. A. Fisher. The use of multiple measurements in taxonomic problems. *Annals of Eugenics*, 7:179–188, 1936.
5. R. W. Hamming. Error detecting and error correcting codes. *The Bell System Technical Journal*, 26(2):147–160, April 1950.
6. V. I. Levenshtein. Binary codes capable of correcting deletions, insertions and reversals. *Soviet Physics Doklady*, 10(8):707–710, 1966.
7. S. S. Stevens. On the theory of scales of measurement. *Science*, 103(2684):677–680, 1946.

Data Preprocessing

3

Abstract

In real world applications, data usually contain errors and noise, need to be scaled and transformed, or need to be collected from different and possibly heterogeneous information sources. We distinguish deterministic and stochastic errors. Deterministic errors can sometimes be easily corrected. Inliers and outliers may be identified and removed or corrected. Inliers, outliers, or noise can be reduced by filtering. We distinguish many different filtering methods with different effectiveness and computational complexities: moving statistical measures, discrete linear filters, finite impulse response, infinite impulse response. Data features with different ranges often need to be standardized or transformed.

3.1 Error Types

Data often contain errors that may cause incorrect data analysis results. We distinguish *stochastic* and *deterministic errors*. Examples for stochastic errors are measurement or transmission errors, which can be modeled by *additive noise*. The left view of Fig. 3.1 shows again the data set from the right view of Fig. 2.5. To mimic a corresponding data set containing stochastic errors we generate *Gaussian noise* data using a random generator that produces data following a Gaussian distribution with mean zero and standard deviation 0.1, a so-called $N(0, 0.1)$ distribution. The middle and right views of Fig. 3.1 show the Gaussian noise data and the data obtained by adding the noise to the original data, respectively. The original (left) and noisy (right) data look very similar, and in fact low noise has often only little impact on data analysis results.

Other types of errors are *inliers* and *outliers*, which are defined as errors in individual data values, where inliers are values that are inside the normal feature distribution, and outliers are outside. Inliers and outliers may be caused by stochastic or deterministic

© Springer Fachmedien Wiesbaden GmbH, part of Springer Nature 2020
T. A. Runkler, *Data Analytics*, https://doi.org/10.1007/978-3-658-29779-4_3

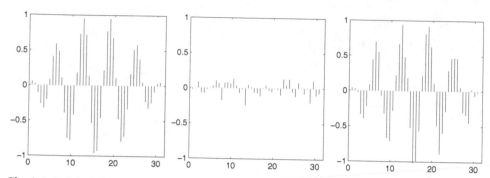

Fig. 3.1 Original data, Gaussian noise, and noisy data

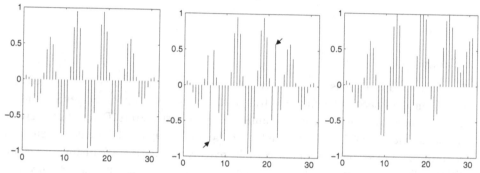

Fig. 3.2 Original data, inliers, and drift

effects, for example by extreme individual measurement errors, or by packet losses in data transmission. In manual data assessment they may be caused when individual data are stored in the wrong data fields, by typos, for example when the decimal point is put at the wrong position. Decimal point errors may also be caused by deterministic effects, for example when data are exchanged between systems using different meanings for the . and , characters, so 1.234 might be transformed to 1,234, which may refer to $1.234 \cdot 10^0$ and $1.234 \cdot 10^3$, depending on country specific notation, and therefore differ by a factor of 1000.

Other types of deterministic errors include the use of wrong formulas for the computation of derived data, or measurement errors caused by wrong calibration, wrong scaling, or sensor drift. Data with such deterministic errors can sometimes be corrected, if the error systematic is known.

Figure 3.2 shows the data set from above again (left), the same data with two inliers (middle), and distorted by a drift effect (right). It is not easy to see the two inliers in the middle view of Fig. 3.2 (marked with little arrows), even though they strongly deviate from the adjacent values, but they are inside the normal feature distribution. However, inliers and outliers may have a strong impact on the data analysis results, so inlier and outlier handling is an important issue in data preprocessing.

However, outliers should be considered with caution, because sometimes data contain unusual but correct data that represent valuable information and should therefore not be considered as errors. For example, if a production plant produces a product with an unusually high or low quality, then the corresponding production data might be considered as outliers, although they might be highly important to identify the influence of production parameters on the product quality. Another example for true anomalies are earthquake events in seismographic records. To distinguish such exotic but valuable data from erroneous data it is often necessary to consult with domain experts.

Outlier detection is easier than inlier detection. A simple method to identify outliers uses feature limits $x_{\min}^{(i)}$ and $x_{\max}^{(i)}$, $i = 1, \ldots, p$, and classifies a value $x_k^{(i)}$ as outlier if it is outside these limits.

$$\left(x_k^{(i)} < x_{\min}^{(i)} \right) \vee \left(x_k^{(i)} > x_{\max}^{(i)} \right) \tag{3.1}$$

The feature limits may be defined by the sign (price, weight, time), the sensor range, the time interval of considered data, or the range of physically meaningful values. Another method to identify outliers is the *2-sigma rule* which considers the mean and the standard deviation of the feature distribution and classifies a value $x_k^{(i)}$ as outlier if it deviates from the mean by at least twice the standard deviation (sigma).

$$\left| x_k^{(i)} - \bar{x}^{(i)} \right| > 2 \cdot s^{(i)} \tag{3.2}$$

where $\bar{x}^{(i)}$ denotes the mean (2.1) and $s^{(i)}$ denotes the standard deviation of the ith feature.

$$s^{(i)} = \sqrt{\frac{1}{n-1} \sum_{k=1}^{n} (x_k^{(i)} - \bar{x}^{(i)})^2} = \sqrt{\frac{1}{n-1} \left(\sum_{k=1}^{n} \left(x_k^{(i)} \right)^2 - n \left(\bar{x}^{(i)} \right)^2 \right)} \tag{3.3}$$

More generally, *m*-sigma rules with arbitrary values of *m* can be defined accordingly.

Inliers like the ones in the middle view of Fig. 3.2 cannot be identified by outlier detection methods. In time series data, inliers may be detected when they significantly deviate from the adjacent values, so a value may be classified as an inlier if the difference from its neighbors is larger than a threshold. A more common approach to remove inliers from time series is filtering, as described in the following section.

Constant data features may be erroneous or correct.

$$x_k^{(i)} = x_l^{(i)} \quad \forall k, l = 1, \ldots, n \tag{3.4}$$

Such constant features do not contain useful information, but they may cause problems with some data analysis methods and may therefore be removed from the data set.

3.2   Error Handling

Individual problematic data such as inliers, outliers, or missing data can be handled in various ways:

1. invalidity list: the data stay unchanged, but the indices of the invalid data are stored in a separate list that is checked in each data processing step.
2. invalidity value: the outlier datum $x_k^{(i)}$ is replaced by a specific invalidity value $x_k^{(i)} = $ NaN, where NaN stands for *not a number*, so the original value is lost. In the 64 bit IEEE floating number format NaN is defined as $7FFFFFFF.
3. Correction or estimation of $x_k^{(i)}$: Only individual features of individual vectors are corrected (if invalid) or estimated (if missing). This can be done in various ways:
   (a) replacing invalid data by the mean, median, minimum, or maximum of the valid feature data x^i.
   (b) nearest neighbor correction: Set $x_k^{(i)} = x_j^{(i)}$ with

   $$\|x_j - x_k\|_{\neg i} = \min_{l \in \{1,...,n\}} \|x_l - x_k\|_{\neg i} \qquad (3.5)$$

   where $\|.\|_{\neg i}$ ignores feature i and invalid or missing data.
   (c) linear interpolation for equidistant time series

   $$x_k^{(i)} = \frac{x_{k-1}^{(i)} + x_{k+1}^{(i)}}{2} \qquad (3.6)$$

   (d) linear interpolation for non-equidistant time series

   $$x_k^{(i)} = \frac{x_{k-1}^{(i)} \cdot (t_{k+1} - t_k) + x_{k+1}^{(i)} \cdot (t_k - t_{k-1})}{t_{k+1} - t_{k-1}} \qquad (3.7)$$

   (e) nonlinear interpolation, for example using splines [1].
   (f) model-based estimation by regression (see Chap. 6).
   (g) filtering (see next section).
4. removal of feature vectors: Each feature vector that contains at least one invalid feature value is removed.
5. removal of features: Each feature that contains at least one invalid feature value is removed.

The choice of the most suitable method depends on the number of available data and on the relative percentage of the invalid data. If only few data are available, and assessment of additional data is difficult or even impossible, then it is often worth the effort to estimate missing data and to correct invalid data. If sufficient data are available and data quality is important, then suspicious data may be completely removed.

3.3   Filtering

Using the methods discussed in the previous section outliers are removed by changing only individual data. The filtering methods presented in this section specifically consider sequential data, and they typically change *all* values of the sequences. The goal is not only to remove inliers and outliers but also to remove noise. Figure 3.3 shows a categorization of the different filter types presented here.

A widely used class of filters uses statistical measures over moving windows. To compute the filtering result for each value x_k, $k = 1, \ldots, n$, all series values in a local window around x_k are considered, and the filter output y_k is the value of a statistical measure of the data in this window. Symmetric windows of the even order $q \in \{2, 4, 6, \ldots\}$ consider the window $w_{kq} = \{x_i \mid i = k - q/2, \ldots, k + q/2\}$ which contains x_k, the $q/2$ previous values and the $q/2$ following values. Symmetric windows are only suitable for offline filtering when the future values of the series are already known. Asymmetric windows of the order $q \in \{0, 1, 2, \ldots\}$ consider the window $w_{kq} = \{x_i \mid i = k - q, \ldots, k\}$ which contains x_k and the q previous values. Asymmetric windows are also suitable for online filtering and are able to provide each filter output y_k as soon as x_k is known.

The mean value is often used as the statistical measure for the data in the window. This yields the symmetric (3.8) and asymmetric (3.9) *moving mean* or *moving average* of the

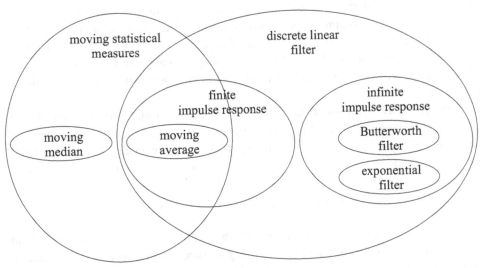

Fig. 3.3 Some important filtering methods for series data

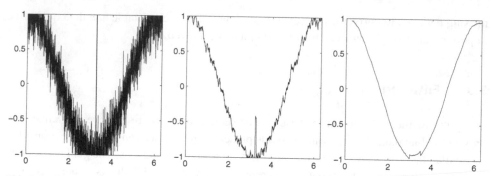

Fig. 3.4 Original data and moving average filtered data, $q = 20$ and $q = 200$

order q defined as

$$y_k = \frac{1}{q+1} \sum_{i=k-\frac{q}{2}}^{k+\frac{q}{2}} x_i \tag{3.8}$$

$$y_k = \frac{1}{q+1} \sum_{i=k-q}^{k} x_i \tag{3.9}$$

The left view of Fig. 3.4 shows a time series $X = \{x_1, \ldots, x_n\}$ generated from cosine data with additive noise and one single peak representing an inlier or outlier. The middle and right views of Fig. 3.4 show the output of the asymmetric moving average filter for $q = 20$ (middle) and $q = 200$ (right). In both filtered time series the noise is substantially reduced. For $q = 200$ the noise is almost completely eliminated. The amplitude of the single peak is reduced from 2 to about 0.5 ($q = 20$) and 0.1 ($q = 200$). Better filter effects can be achieved by larger values of the window size q, but $q - 1$ data points are lost by filtering. Therefore, the window size should be much smaller than the length of the time series to be filtered, $q \ll n$.

In the previous chapter we discussed the median as a statistical measure for ordinal, interval, and ratio scaled data. Using the median as the statistical measure over symmetric or asymmetric moving windows yields the symmetric and asymmetric *moving median* of the order q defined as $m_{kq} \in w_{kq} = \{x_{k-q/2}, \ldots, x_{k+q/2}\}$ or $w_{kq} = \{x_{k-q}, \ldots, x_k\}$ with

$$|\{x_i \in w_{kq} \mid x_i < m_{kq}\}| = |\{x_i \in w_{kq} \mid x_i > m_{kq}\}| \tag{3.10}$$

where $|.|$ denotes the set cardinality. Figure 3.5 shows the same situation as in Fig. 3.4, but with the moving median instead of the moving average. For the same window size, the moving average and the moving median achieve similar degrees of noise reduction, but the moving median achieves a much better suppression of the peak.

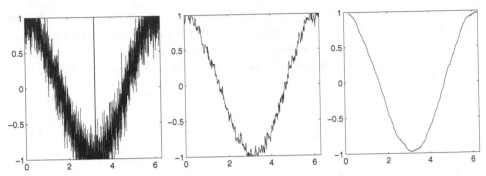

Fig. 3.5 Original data and moving median filtered data, $q = 20$ and $q = 200$

The third family of filtering methods presented here is the *exponential filter*. The exponential filter works best with slow changes of the filtered data, so each value of the filter output y_k is similar to the previous value of the filter output y_{k-1}, except for a correction term that is computed as a fraction $\eta \in [0, 1]$ of the difference between the current input x_k and previous output y_{k-1}. So, the exponential filter is defined as

$$y_k = y_{k-1} + \eta \cdot (x_k - y_{k-1}), \quad k = 1, \ldots, n - 1 \tag{3.11}$$

with an initial filter output $y_0 = 0$. The current filter output y_k is affected by each past filter output y_{k-i}, $i = 1, \ldots, k - 1$, with the multiplier $(1 - \eta)^i$, so the filter exponentially forgets previous filter outputs, hence the name exponential filter. For $\eta = 0$ the exponential filter maintains the initial value $y_k = y_0 = 0$. For $\eta = 1$ it yields the current filter input, $y_k = x_k$. So, for nontrivial filter behavior η should be chosen larger than zero but smaller than one. Figure 3.6 shows the output of the exponential filter for the example data set in Figs. 3.4 and 3.5, with the parameter values $\eta = 0.1, 0.01$, and 0.001. The lower the value of η, the higher the noise reduction. The reduction of the peak is much weaker than with the median filter. Notice the lag for smaller values of η. For $\eta = 0.001$ the filter output is not able to follow the amplitude of the original data any more. So, for the exponential filter

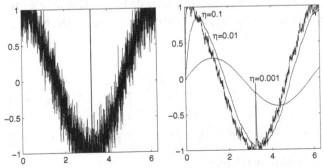

Fig. 3.6 Original data and exponentially filtered data, $\eta = 0.1, 0.01$, and 0.001

the parameter η has to be chosen carefully. It has to be small enough to achieve a sufficient filter effect but large enough to maintain the essential characteristics of the original data.

The moving average and the exponential filter are instances of the more general family of *discrete linear filters*. An asymmetric discrete linear filter of the order $q = 0, 1, 2, \ldots$ is defined by

$$\sum_{i=0}^{q} a_i \cdot y_{k-i} = \sum_{i=0}^{q} b_i \cdot x_{k-i} \tag{3.12}$$

with the filter coefficients $a_0, \ldots, a_q, b_0, \ldots, b_q \in \mathbb{R}$ [3]. After some simple conversions we obtain the output of an asymmetric discrete linear filter as

$$y_k = \sum_{i=0}^{q} \frac{b_i}{a_0} \cdot x_{k-i} - \sum_{i=1}^{q} \frac{a_i}{a_0} \cdot y_{k-i} \tag{3.13}$$

For simplicity we consider only the asymmetric case here. The reader may easily modify the indices to obtain the equations for the symmetric discrete linear filter. The properties of a discrete linear filter are specified by the coefficient vectors $a = (a_0, \ldots, a_q)$ and $b = (b_0, \ldots, b_q)$, where $a_0 \neq 0$. If $a_1 = \ldots = a_q = 0$, then the filter output y_k only depends on the inputs $x_{k-q}, \ldots, x_k$, and is independent of the previous outputs $y_{k-q}, \ldots, y_{k-1}$, so a change in the input x_k affects only the current output y_k and the following outputs $y_{k+1}, \ldots, y_{k+q}$, and is then completely forgotten. For $a_1 = \ldots = a_q = 0$, a discrete linear filter is therefore called a *finite impulse response (FIR) filter*. Otherwise, we call the discrete linear filter an *infinite impulse response (IIR) filter*. Figure 3.7 shows the data flow graph of an FIR filter. Each circle with a plus sign (+) represents the addition of the node inputs, each labeled edge represents multiplication by the values of the edge label, and each empty circle represents a time delay of one step, i.e. in each step the current value is stored and the previous value is retrieved (a so-called *register*). The FIR data flow graph is a sequence of q stages, where each stage requires a multiplication, addition, and storage. Figure 3.8 shows the data flow graph of an IIR filter. Each stage of the IIR data flow graph requires two multiplications, an addition of three arguments, and the storage of two values. So-called *signal processors* provide dedicated hardware units that are able to compute individual FIR and IIR stages in only a few clock cycles, which enables very fast filtering.

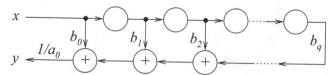

Fig. 3.7 Data flow graph of an FIR filter

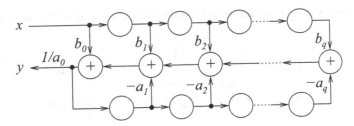

Fig. 3.8 Data flow graph of an IIR filter

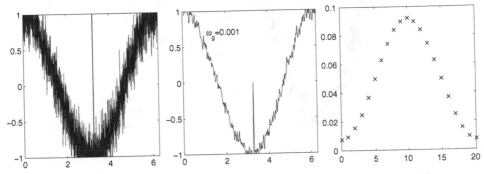

Fig. 3.9 Original data, FIR low pass filter (order 20) filtered data, and FIR filter coefficients

The moving average and exponential filter are instances of discrete linear filters. Inserting the coefficient vectors $a = (1, \eta - 1), b = (\eta, 0)$ into (3.13) yields

$$y_k = \eta \cdot x_k - (\eta - 1) \cdot y_{k-1} = y_{k-1} + \eta \cdot (x_k - y_{k-1}) \tag{3.14}$$

which is equal to the exponential filter at (3.11). Inserting the coefficient vectors $a = (1)$, $b = (\underbrace{\dfrac{1}{q+1}, \ldots, \dfrac{1}{q+1}}_{q+1 \text{ times}})$ into (3.13) yields

$$y_k = \sum_{i=0}^{q} \frac{1}{q+1} \cdot x_{k-i} = \frac{1}{q+1} \sum_{i=k-q}^{k} x_i \tag{3.15}$$

which is equal to the (asymmetric) moving average filter at (3.9). The moving average filter of order $q = 2$ is also called a *first order FIR low pass filter* which means that it lets the low frequencies pass. Many other types of low pass, high pass, and other filter types have been defined. FIR or IIR filters of specific types with desired properties can be designed using filter coefficient tables or software tools for filter design.

Figure 3.9 shows the example data set from Figs. 3.4, 3.5, and 3.6 (left), the output of an FIR low pass filter of order 20 (middle), and the corresponding filter parameters $b_0, \ldots, b_{20}$ (right). All filter coefficients b are positive and are on a symmetric bell-shaped

Table 3.1 Filter coefficients for the second order Butterworth low pass filter for different limit frequencies

ω_g	a_0	a_1	a_2	b_0	b_1	b_2
0.01	1	-1.96	0.957	$2.41 \cdot 10^{-4}$	$4.83 \cdot 10^{-4}$	$2.41 \cdot 10^{-4}$
0.003	1	-1.99	0.987	$2.21 \cdot 10^{-5}$	$4.41 \cdot 10^{-5}$	$2.21 \cdot 10^{-5}$
0.001	1	-2	0.996	$2.46 \cdot 10^{-6}$	$4.92 \cdot 10^{-6}$	$2.46 \cdot 10^{-6}$

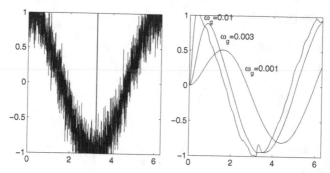

Fig. 3.10 Original data and second order Butterworth low pass filtered data, $\omega_g = 0.01, 0.003$, and 0.001

curve. The largest value has the middle coefficient b_{10}, and the sum of all coefficients is $b_0 + \ldots + b_{20} = 1$. As for all FIR filters, we have $a = (1)$. The output of the FIR low pass filter (middle view) is similar to the output of the moving average filter at Fig. 3.4.

A popular family of IIR filters are the so-called Butterworth filters [2]. A first order Butterworth low pass filter with limit frequency $\omega_g = 0.5$ has the filter coefficients $a = (1)$, $b = (0.5, 0.5)$, and corresponds to the first order FIR low pass filter and the second order moving average filter. Table 3.1 shows the filter coefficients for the second order Butterworth low pass filter for the limit frequencies $\omega_g = 0.01, 0.003$, and 0.001. For $\omega_g \to 0$ we obtain $a = (1, -2, 1)$ and $b = (0, 0, 0)$. This corresponds to an autoregressive system (see Sect. 7.3) with

$$y_k = -2y_{k-1} + y_{k-2} \tag{3.16}$$

which for zero initialization $y_0 = y_1 = 0$ yields constant zero outputs $y_k = 0$, and is unstable for all other initializations. Figure 3.10 shows the output of second order Butterworth low pass filters with the parameter values from Table 3.1. For $\omega_g = 0.01$, noise and also the peak are considerably reduced. Smaller limit frequencies cause the same effect that we already observed for exponential filters (Fig. 3.6): The filter output lags the filter input more and more as ω_g decreases, until for $\omega_g \to 0$ the output becomes zero. These IIR low pass filters have only 6 parameters, as opposed to the 22 parameters of the FIR low pass filter from Fig. 3.9. Compared to FIR filters, IIR filters need less parameters and require a lower computational effort, but are more sensitive to changes in their parameters and may become unstable.

3.4   Data Transformation

Different features may have considerably different ranges. For example, the price of a car in Euros and its horse power may differ by several orders of magnitude. If such features are used together, incorrect results may be obtained because the ranges of the features are so different. The left view of Fig. 3.11 shows a data set which is randomly generated using two-dimensional Gaussian distribution with mean $\mu = (30000, 100)$ and standard deviation $s = (9000, 30)$, so the two features are uncorrelated and differ by a factor of 300. The data set appears as a horizontal line close to zero, so only the first (larger) feature is visible. This visualization might be appropriate, for example if the features are the lengths and widths of steel tubes in millimeters. But it might be inappropriate if the features are the price (in Euros) and horse power of cars, which might be considered equally important features. Also the choice of the feature units might be arbitrary. In our example it should not matter if we measure the car's power in horse power, kilowatts, or watts. In such cases it is useful to transform the feature data to similar data ranges. This transformation is called *standardization*. The *minimal hypercube* of a p-dimensional data set X is defined as

$$H(X) = [\min\{X^{(1)}\}, \max\{X^{(1)}\}] \times \ldots \times [\min\{X^{(p)}\}, \max\{X^{(p)}\}] \qquad (3.17)$$

so $H(X)$ contains all points in X, or $X \subseteq H$. If X contains outliers or if it only partially covers the actually relevant feature space, then it might be more suitable not to consider the *observed* minimal hypercube but the *relevant hypercube*

$$H^*(X) = [x_{\min}^{(1)}, x_{\max}^{(1)}] \times \ldots \times [x_{\min}^{(p)}, x_{\max}^{(p)}] \qquad (3.18)$$

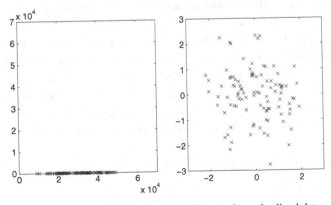

Fig. 3.11 Two features with different ranges: original data and standardized data

where the limits x_{min} and x_{max} might come from X, so $x_{min} = \min\{X^{(i)}\}$ or $x_{max} = \max\{X^{(i)}\}$, or might be arbitrarily defined, just as the feature limits in the outlier detection at (3.1). For standardization the observed hypercube is mapped to the unit hypercube, i.e. the values of each feature are mapped to the unit interval $[0, 1]$ by

$$y_k^{(i)} = \frac{x_k^{(i)} - x_{min}^{(i)}}{x_{max}^{(i)} - x_{min}^{(i)}} \qquad (3.19)$$

which is called *hypercube standardization*. If outliers are not removed before hypercube standardization with feature limits, then they will be mapped to values outside the unit interval $[0, 1]$. If standardization is done using the observed hypercube, then it is important to remove outliers before standardization because they may otherwise strongly affect feature minima or maxima.

Alternative standardization methods use statistical transformations of the data set X, for example using the mean $\bar{x}$ (2.1) and the standard deviation s_x (3.3) of each feature, which leads to the so-called μ-σ-*standardization*

$$y_k^{(i)} = \frac{x_k^{(i)} - \bar{x}^{(i)}}{s_x^{(i)}} \qquad (3.20)$$

The right view of Fig. 3.11 shows the μ-σ-standardization of the data set shown on the left. The standardized data do not appear as a line but as a point cloud that possibly reflects the data structure more appropriately, but this depends on the semantics of the data, as pointed out before. Also for μ-σ-standardization outliers should be remove before, they may otherwise strongly affect feature means or standard deviations.

Hypercube standardization is appropriate for features distributed more or less uniformly across intervals with finite lower and upper limits, and μ-σ-standardization is appropriate for features with approximate Gaussian distributions. Often features do not correspond to either of these types of distributions. For example, time differences between random events like order arrivals may follow a Poisson distribution that always yields positive values (lower bound zero) but allows arbitrarily large values with low probabilities (upper bound infinity). Hypercube or μ-σ-standardization would not yield reasonable representations of such features. Instead, such features are often first transformed such that the resulting distribution can be better approximated by uniform or Gaussian distributions. To choose an appropriate transformation it is useful to first consider the observed and the desired data range (unlimited, lower limit, upper limit, or two limits). The following list presents some frequently used data transformations and the corresponding data ranges.

- inverse transformation $f : \mathbb{R}\backslash\{0\} \to \mathbb{R}\backslash\{0\}$

$$f(x) = f^{-1}(x) = \frac{1}{x} \qquad (3.21)$$

- root transformation $f : (c, \infty) \rightarrow \mathbb{R}^+$

$$f(x) = \sqrt[b]{x - c} \tag{3.22}$$

$$f^{-1}(x) = x^b + c, \quad c \in \mathbb{R},\ b > 0 \tag{3.23}$$

- logarithmic transformation $f : (c, \infty) \rightarrow \mathbb{R}$

$$f(x) = \log_b(x - c) \tag{3.24}$$

$$f^{-1}(x) = b^x + c, \quad c \in \mathbb{R},\ b > 0 \tag{3.25}$$

- Fisher-Z transformation $f : (-1, 1) \rightarrow \mathbb{R}$

$$f(x) = \text{artanh}\, x = \frac{1}{2} \cdot \ln \frac{1 + x}{1 - x} \tag{3.26}$$

$$f^{-1}(x) = \tanh x = \frac{e^x - e^{-x}}{e^x + e^{-x}} \tag{3.27}$$

3.5 Data Integration

In many data analytics projects the relevant data are not contained in a single data set but come from different data sets, files, data bases, or systems. For example, prices, sales figures, logistics data, and manufacturing parameters might be stored in different data sets and need to be merged. This means that feature vectors from different data sets have to be assigned to each other. This assignment can be done based on specific *label features* like codes of people or objects, (relative) time stamps, or (relative) locations. Figure 3.12 shows the scheme of label based integration of data sets. Feature vectors with the same labels are concatenated. Suitable mechanisms need to be defined if the labels only match approximately, for example, two time stamps 10:59 and 11:00 might be considered equivalent. Missing data might be generated if a label in one data set does not match labels in all other data sets. Sometimes feature vectors with such missing data are removed. If multiple data are found for the same label, multiple feature vectors may be generated, or the feature vectors may be combined to a single feature vector, for example by averaging.

Fig. 3.12 Label based integration of data sets

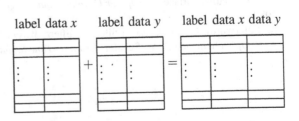

Data transformation and data integration are part of the so-called *ETL process* (extraction, transformation, loading). ETL processes may be designed and implemented using specific software tools.

Problems

3.1 Consider the time series $(920, 980, 1.03, 950, 990)$.

(a) Which stochastic and deterministic errors does this time series seem to contain?
(b) What may be reasons for these errors?
(c) Compute the output of an asymmetric median filter with window length 3 for this time series.
(d) Which effect does this filter have on the observed errors?

3.2 Which of these filters are FIR, IIR, or none of these?

(a) $x_k + x_{k-1} + y_k = 0$
(b) $x_k + x_{k-1} + x_{k-2} = 0$
(c) $x_k + y_{k-1} + y_k = 0$

3.3 Consider the IIR filter $y_k = 2y_{k-1} - y_{k-1} + x_k + x_{k-1}, k = 3, 4, \ldots, y_1 = y_2 = 0$.

(a) What is the filter output sequence y for the input sequence $x = (0, 0, 1, 0, 0, 0, 0, 0)$?
(b) What is the filter output sequence y for the input sequence $x = (0, 0, 1, a, b, 0, 0, 0)$, $a, b \in \mathbb{R}$?
(c) Give a formula for the filter output y_k, $k = 8, 9, 10, \ldots$, for $x = (0, 0, 1, a, b, 0, 0, 0, \ldots)$, $a, b \in \mathbb{R}$!
(d) For which finite values of a and b will the filter be unstable?
(e) For which finite values of a and b will the filter converge to $\lim_{k \to \infty} y_k = 0$?

References

1. B. A. Barsky and D. P. Greenberg. Determining a set of B–spline control vertices to generate an interpolating surface. *Computer Graphics and Image Processing*, 14(3):203–226, November 1980.
2. S. Butterworth. On the theory of filter amplifiers. *Wireless Engineer*, 7:536–541, 1930.
3. A. V. Oppenheim and R. W. Schafer. *Discrete–Time Signal Processing*. Prentice Hall, 2009.

Data Visualization

4

Abstract

Data can often be very effectively analyzed using visualization techniques. Standard visualization methods for object data are plots and scatter plots. To visualize high-dimensional data, projection methods are necessary. We present linear projection (principal component analysis, Karhunen-Loève transform, singular value decomposition, eigenvector projection, Hotelling transform, proper orthogonal decomposition, multidimensional scaling) and nonlinear projection methods (Sammon mapping, auto-encoder). Data distributions can be estimated and visualized using histogram techniques. Periodic data (such as time series) can be analyzed and visualized using spectral analysis (cosine and sine transforms, amplitude and phase spectra).

4.1   Diagrams

The human eye (and brain) is a very powerful tool for analyzing data. Therefore, data visualization plays an important role in data analysis [2, 6, 13]. Visualized data are also useful for documenting and communicating data analysis results, for example for discussions with domain experts. Paper and screens allow for two-dimensional visualization. Projection methods are required for the visualization of higher-dimensional data. Figure 4.1 shows the visualization and projection methods presented in this chapter.

Two-dimensional visualization uses two orthogonal coordinate axes and represents each feature vector as a point in the coordinate system. The visualization of only one feature is called a (simple) *diagram*. Visualizations of more than one feature can be done with *scatter diagrams*. A two-dimensional scatter diagram matches each feature to one of the two coordinate axes, so the feature plane matches the visualization plane. Higher-dimensional scatter diagrams project the higher dimensional feature space to the two-dimensional visualization plane or use specific symbols (geometric characters, numbers, grey values, or

© Springer Fachmedien Wiesbaden GmbH, part of Springer Nature 2020
T. A. Runkler, *Data Analytics*, https://doi.org/10.1007/978-3-658-29779-4_4

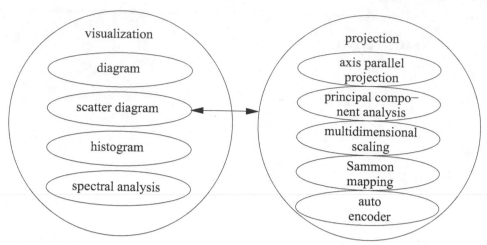

Fig. 4.1 Visualization and projection methods

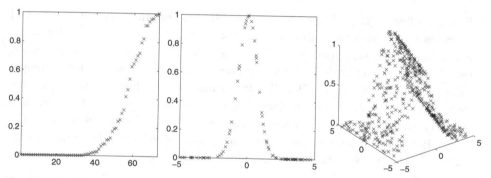

Fig. 4.2 Simple diagram, two-dimensional and three-dimensional scatter diagram

colors) to represent specific features. Three-dimensional scatter diagrams often use linear projections of the three-dimensional feature space to the two-dimensional visualization plane. Consider a three-dimensional data set $X = \{(x_1, y_1, z_1), \ldots, (x_n, y_n, z_n)\}$. Figure 4.2 shows a simple diagram $X^{(1)} = \{x_1, \ldots, x_n\}$ (left), a two-dimensional scatter diagram $X^{(1,2)} = \{(x_1, y_1), \ldots, (x_n, y_n)\}$ (middle), and a three-dimensional scatter diagram $X^{(1,\ldots,3)} = \{(x_1, y_1, z_1), \ldots, (x_n, y_n, z_n)\}$ (right). The simple diagram shows only the first feature, the two-dimensional scatter diagram displays the first and the second features, and the three-dimensional scatter diagram displays all three features. A three-dimensional scatter plot is only useful if the viewpoint is chosen appropriately. The simplest kind of projections are axis-parallel projections. In Chap. 6 we will present feature selection methods which yield axis-parallel projections. Figure 4.3 shows the three two-dimensional axis-parallel projections (x, y), (x, z), and (y, z) of the three-dimensional data set from above. Such axis-parallel projections can be simply produced by omitting the features x, y, or z. The two-dimensional diagrams then display the remaining pairs

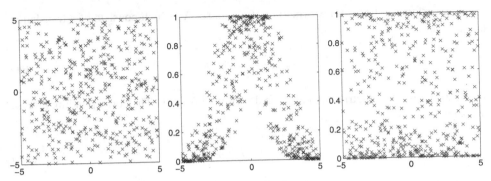

Fig. 4.3 The three two-dimensional axis-parallel projections

of features. In this example, the axis-parallel projections do not allow us to appropriately assess the geometric relation between the three features. In the next sections we present more appropriate linear and nonlinear projection methods for visualization that allow to better assess the structure of high-dimensional data sets.

4.2 Principal Component Analysis

The *principal component analysis (PCA)* [7] is also called *Karhunen-Loève transform, singular value decomposition (SVD), eigenvector projection, Hotelling transform,* or *proper orthogonal decomposition.* The main idea of PCA is to find a linear projection of the data that optimally matches the data structure in the well defined sense of accounting for the maximum amount of variance that can be captured in the lower dimensional representations of the data. Figure 4.4 shows a two-dimensional data set. A projection of these data to the axes y_1 and y_2 maximizes the amount of original variance that can be captured by any linear projection. The vectors y_1 and y_2 are called the *main axes* or *principal components* of the data set, hence the name principal component analysis.

Fig. 4.4 Principal component analysis

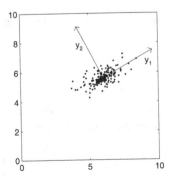

PCA performs a linear transformation of a data set X, which consists of a translation and a rotation.

$$y_k = (x_k - \bar{x}) \cdot E \tag{4.1}$$

where E is a rotation matrix that has to be determined from X. The corresponding inverse transformation is

$$x_k = y_k \cdot E^T + \bar{x} \tag{4.2}$$

To determine the rotation matrix E, the variance of Y is maximized. The variance of Y can be written as

$$v_y = \frac{1}{n-1} \sum_{k=1}^{n} y_k^T y_k \tag{4.3}$$

$$= \frac{1}{n-1} \sum_{k=1}^{n} \left((x_k - \bar{x}) \cdot E \right)^T \cdot \left((x_k - \bar{x}) \cdot E \right) \tag{4.4}$$

$$= \frac{1}{n-1} \sum_{k=1}^{n} E^T \cdot (x_k - \bar{x})^T \cdot (x_k - \bar{x}) \cdot E \tag{4.5}$$

$$= E^T \left(\frac{1}{n-1} \sum_{k=1}^{n} (x_k - \bar{x})^T \cdot (x_k - \bar{x}) \right) \cdot E \tag{4.6}$$

$$= E^T \cdot C \cdot E \tag{4.7}$$

where C is the covariance matrix of X. The elements of the covariance matrix are

$$c_{ij} = \frac{1}{n-1} \sum_{k=1}^{n} (x_k^{(i)} - \bar{x}^{(i)})(x_k^{(j)} - \bar{x}^{(j)}), \quad i, j = 1, \ldots, p \tag{4.8}$$

The transformation matrix E should only represent a rotation, not a dilation, so we require

$$E^T \cdot E = 1 \tag{4.9}$$

This leads to a constrained optimization problem that can be solved using Lagrange optimization. For details on Lagrange optimization please refer to the Appendix. The variance (4.7) can be maximized under the constraint (4.9) using the Lagrange function

$$L = E^T C E - \lambda (E^T E - 1) \tag{4.10}$$

The necessary condition for optima of L is

$$\frac{\partial L}{\partial E} = 0 \quad \Rightarrow \quad CE - \lambda E = 0 \quad \Rightarrow \quad CE = \lambda E \qquad (4.11)$$

which defines an eigenproblem that can be solved, for example, by conversion into the homogeneous equation system

$$(C - \lambda I) \cdot E = 0 \qquad (4.12)$$

The rotation matrix E is the concatenation of the eigenvectors of C.

$$E = (v_1, \ldots, v_p), \quad (v_1, \ldots, v_p, \lambda_1, \ldots, \lambda_p) = \text{eig } C \qquad (4.13)$$

The variances in Y correspond to the eigenvalues $\lambda_1, \ldots, \lambda_p$ of C because

$$CE = \lambda E \quad \Rightarrow \quad \lambda = E^T C E = v_y \qquad (4.14)$$

This means that PCA not only yields the coordinate axes in Y and the transformation matrix that maximizes the variances in Y but also the *values* of these variances. These can be used to select the projection dimensions in Y, so that the variances are maximized. To map a data set $X \subset \mathbb{R}^p$ to a data set $Y \subset \mathbb{R}^q$ with $1 \leq q < p$, we use the rotation and projection matrix E generated by concatenating the eigenvectors with the q largest eigenvalues (corresponding to the q dimensions with the largest variance).

$$E = (v_1, \ldots, v_q) \qquad (4.15)$$

In data visualization the projection dimension is usually $q = 2$, so the high-dimensional data are rotated and mapped to a two dimensional scatter plot. In other cases, dimension reduction may be done for efficiency reasons, for example when the time or space complexity of a data analysis algorithm increases strongly with the dimensionality of the data. In such cases, the goal is to reduce the number of dimensions as much as possible but to keep a certain percentage of the data variance, say 95%. In this case we choose the minimum possible q so that

$$\sum_{i=1}^{q} \lambda_i \Big/ \sum_{i=1}^{p} \lambda_i \geq 95\% \qquad (4.16)$$

For $p = q$, PCA performs a translation and a rotation of the data set, no projection, so the inverse PCA completely recovers the original data and the transformation error is zero. For $q < p$, PCA performs a projection that causes an information loss, so inverse PCA yields x'_k, $k = 1, \ldots, n$, with typically $x'_k \neq x_k$. It can be shown that the average

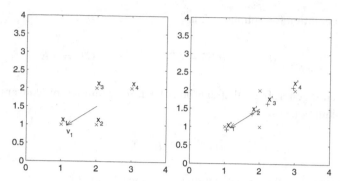

Fig. 4.5 Principal component analysis (four points data set)

quadratic transformation error is proportional to the sum of the eigenvalues of the omitted eigenvectors.

$$e = \frac{1}{n} \sum_{k=1}^{n} \|x_k - x'_k\|^2 = \frac{n-1}{n} \sum_{i=q+1}^{p} \lambda_i \qquad (4.17)$$

PCA chooses the eigenvectors with the largest eigenvalues, so PCA yields the projection with not only the largest variance but also the smallest quadratic transformation error. In our derivation of PCA we maximized the variance and found that the same method minimizes the quadratic transformation error. PCA can also be derived in the reverse order: if we minimize the quadratic transformation error, then we will find that the same method maximizes the variance.

We illustrate PCA with a simple example. Figure 4.5 shows the data set

$$X = \{(1, 1), (2, 1), (2, 2), (3, 2)\} \qquad (4.18)$$

with mean (2.1) and covariance (4.8)

$$\bar{x} = \frac{1}{2} \cdot (4, 3) \qquad (4.19)$$

$$C = \frac{1}{3} \cdot \begin{pmatrix} 2 & 1 \\ 1 & 1 \end{pmatrix} \qquad (4.20)$$

and the eigenvalues and eigenvectors (4.13)

$$\lambda_1 = 0.8727 \qquad (4.21)$$

$$\lambda_2 = 0.1273 \qquad (4.22)$$

$$v_1 = \begin{pmatrix} -0.85065 \\ -0.52573 \end{pmatrix} \tag{4.23}$$

$$v_2 = \begin{pmatrix} 0.52573 \\ -0.85065 \end{pmatrix} \tag{4.24}$$

If we map the $p = 2$ dimensional data set to a $q = 1$ dimensional data set, then we choose v_1, the eigenvector with the largest eigenvalue, for the rotation matrix. This projection covers 87% (4.21) of the variances and has an average quadratic projection error of $(4 - 1)/4 \cdot 12.73\% \approx 9.5\%$ (4.17), (4.22). The new coordinate system consists of only one axis that is shown in Fig. 4.5 as an arrow v_1 originating at $\bar{x}$. The rotation (and projection) matrix is

$$E = v_1 = \begin{pmatrix} -0.85065 \\ -0.52573 \end{pmatrix} \tag{4.25}$$

which yields the projected data (4.1)

$$Y = \{1.1135, 0.2629, -0.2629, -1.1135\} \tag{4.26}$$

Inverse PCA (4.2) yields

$$X' = \{ (1.0528, 0.91459), (1.7764, 1.3618), \\ (2.2236, 1.6382), (2.9472, 2.0854) \} \neq X \tag{4.27}$$

These projections X' are on the principal axis of X, the coordinate vector v_1, see Fig. 4.5 (right).

4.3 Multidimensional Scaling

Multidimensional scaling (MDS) [14] is a linear mapping based on matrix decomposition. Given the data matrix $X \in \mathbb{R}^{n \times p}$, eigendecomposition of the positive semi-definite matrix XX^T yields

$$XX^T = Q\Lambda Q^T = (Q\sqrt{\Lambda}^T) \cdot (\sqrt{\Lambda}Q^T) = (Q\sqrt{\Lambda}^T) \cdot (Q\sqrt{\Lambda}^T)^T \tag{4.28}$$

where $Q = (v_1, \ldots, v_p)$ is the matrix of eigenvectors of XX^T and Λ is the diagonal matrix whose diagonal elements are the corresponding eigenvalues of XX^T, $\Lambda_{ii} = \lambda_i$, $i = 1, \ldots, p$. The rank of XX^T is less or equal to p, where often $p < n$, so XX^T has less than n distinct eigenvalues and eigenvectors. Using eigendecomposition, an estimate for

X is

$$Y = Q\sqrt{\Lambda}^T \tag{4.29}$$

To produce lower dimensional projections $Y \subset \mathbb{R}^q$, $q < p$, only the first q dimensions are used and then scaled so that their squared norms are equal to the corresponding eigenvalues.

MDS of a feature data set X yields the same results as PCA. However, MDS cannot only be used to map a feature data set X to a lower-dimensional feature data set Y but also to produce an (approximate) feature space representation Y for relational data specified by a distance matrix D. Assume that D is associated with an unknown data set $\tilde{X} \in \mathbb{R}^{n \times p}$, arbitrarily choose $\tilde{x}_a$, $a \in \{1, \ldots, n\}$, as an *anchor point*, and transform $\tilde{X}$ to X in a coordinate system with origin $\tilde{x}_a$, so

$$x_k = \tilde{x}_k - \tilde{x}_a \tag{4.30}$$

$k = 1, \ldots, n$, and

$$\tilde{x}_i - \tilde{x}_j = x_i - x_j \tag{4.31}$$

$i, j = 1, \ldots, n$. Taking the scalar product of each side with itself yields

$$(\tilde{x}_i - \tilde{x}_j)(\tilde{x}_i - \tilde{x}_j)^T = (x_i - x_j)(x_i - x_j)^T \tag{4.32}$$

$$\Rightarrow d_{ij}^2 = x_i x_i^T - 2 x_i x_j^T + x_j x_j^T = d_{ia}^2 - 2 x_i x_j^T + d_{ja}^2 \tag{4.33}$$

$$\Rightarrow x_i x_j^T = (d_{ia}^2 + d_{ja}^2 - d_{ij}^2)/2 \tag{4.34}$$

so each element of the product matrix XX^T can be computed from the elements of the distance matrix D. If D is an Euclidean distance matrix, then each anchor $a = 1, \ldots, n$ will yield the same XX^T. Otherwise XX^T may be averaged over several or all anchors.

We illustrate MDS with the example data set from the previous section, so consider again the data set from Fig. 4.5. Subtracting the mean yields

$$X = \left\{ \left(-1, -\frac{1}{2}\right), \left(0, -\frac{1}{2}\right), \left(0, \frac{1}{2}\right), \left(1, \frac{1}{2}\right) \right\} \tag{4.35}$$

and the product matrix

$$XX^T = \frac{1}{4} \begin{pmatrix} 5 & 1 & -1 & -5 \\ 1 & 1 & -1 & -1 \\ -1 & -1 & 1 & 1 \\ -5 & -1 & 1 & 5 \end{pmatrix} \tag{4.36}$$

The largest eigenvalue and the corresponding eigenvector of XX^T are

$$\lambda_1 \approx 2.618 \tag{4.37}$$

$$v_1 \approx \begin{pmatrix} -0.6882 \\ -0.1625 \\ 0.1625 \\ 0.6882 \end{pmatrix} \tag{4.38}$$

and with (4.29) we finally obtain the MDS projection

$$Y \approx \begin{pmatrix} -0.6882 \\ -0.1625 \\ 0.1625 \\ 0.6882 \end{pmatrix} \sqrt{2.618} \approx \begin{pmatrix} -1.1135 \\ -0.2629 \\ 0.2629 \\ 1.1135 \end{pmatrix} \tag{4.39}$$

which is equal to the results obtained by PCA (4.26).

The quality of a projection can be visualized by a so-called *Shepard diagram*, a scatter plot of the distances d_{ij}^y in the projection versus the corresponding distances d_{ij}^x of the original data, $i, j = 1, \ldots, n$. For our example we have

$$D^x = \begin{pmatrix} 0 & 1 & \sqrt{2} & \sqrt{5} \\ 1 & 0 & 1 & \sqrt{2} \\ \sqrt{2} & 1 & 0 & 1 \\ \sqrt{5} & \sqrt{2} & 1 & 0 \end{pmatrix} \approx \begin{pmatrix} 0 & 1 & 1.4142 & 2.2361 \\ 1 & 0 & 1 & 1.4142 \\ 1.4142 & 1 & 0 & 1 \\ 2.2361 & 1.4142 & 1 & 0 \end{pmatrix} \tag{4.40}$$

$$D^y \approx \begin{pmatrix} 0 & 0.8507 & 1.3764 & 2.2270 \\ 0.8507 & 0 & 0.5257 & 1.3764 \\ 1.3764 & 0.5257 & 0 & 0.8507 \\ 2.2270 & 1.3764 & 0.8507 & 0 \end{pmatrix} \tag{4.41}$$

Figure 4.6 shows the corresponding Shepard diagram. Both distance matrices D^x and D^y are symmetric with zero diagonals, so we consider only the upper or lower triangle. Multiple points are displayed as ticks with circles. A zero-error projection would yield

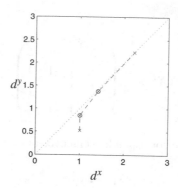

Fig. 4.6 Shepard diagram for PCA/MDS projection (four points data set)

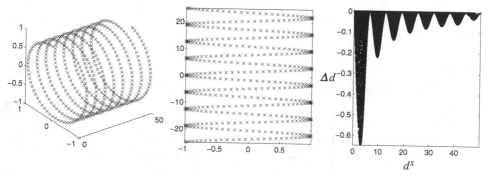

Fig. 4.7 Helix data set, PCA/MDS projection, and projection errors

only points on the main diagonal with $d_{ij}^x = d_{ij}^y$ but this can usually not be achieved. For this example PCA/MDS yields a Shepard diagram with several points very close to the main diagonal but one point is quite distant (referring to $d_{23}^x = d_{32}^x = 1$ and $d_{23}^y = d_{32}^y \approx 0.5257$). So PCA/MDS yields a very accurate projection for many pairwise distances at the expense of one distance that is not mapped well. The points in the Shepard diagram are connected by dashed lines, from bottom to top, with increasing distance d^y. In this example the line segments show that the MDS mapping is monotonic, i.e. increasing d^x does not decrease d^y. Finding such monotonic mappings is the purpose of an alternative method proposed by Torgerson [12].

We further illustrate PCA/MDS using two more complex data sets. The first data set is a 3D helix curve defined by

$$X = \{(t, \sin t, \cos t)^T \mid t \in \{0, 0.1, 0.2, \ldots, 50\}\} \tag{4.42}$$

Figure 4.7 shows this data set (left), its PCA/MDS projection onto the plane (center), and its projection error $\Delta d = d_{ij}^y - d_{ij}^x$ over d_{ij}^x (right). We plot the projection errors Δd here because they are much lower than the distances d^x, so a Shepard diagram would only

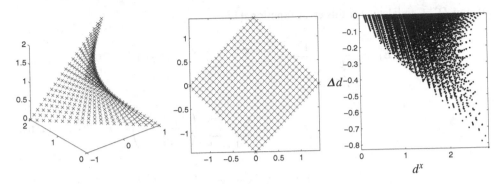

Fig. 4.8 Bent square data set, PCA/MDS projection, and projection errors

display a very close approximate of the main diagonal. The PCA/MDS mapping (Fig. 4.7, center) represents the data structure quite well, and the projection errors are low. As for all linear projection methods, pairwise distances in the projection space can never be larger than the original distances, so all of the PCA/MDS projection errors are non-positive, which leads to a bias towards lower distances d^y, and therefore the average of Δd is not zero but negative.

The second more complex data set considered here is a 3D bent square defined by

$$X = \{((t_1 - 1) \cdot (t_2 - 1), t_1, t_2)^T \mid t_1, t_2 \in \{0, 0.1, 0.2, \ldots, 2\}\} \tag{4.43}$$

Figure 4.8 shows this data set, its PCA/MDS projection onto the plane, and its projection error. The PCA/MDS mapping represents the data structure very well, but again the projection errors have a negative bias.

PCA/MDS is a linear mapping with low computational effort and high robustness. Complicated nonlinear data structures cannot be appropriately projected by linear mapping but require nonlinear approaches. The following two sections present two popular examples for nonlinear projections: Sammon mapping and auto-encoder.

4.4   Sammon Mapping

The idea of Sammon mapping [9] is to map a data set $X \subset \mathbb{R}^p$ to a data set $Y \subset \mathbb{R}^q$ so that distances between pairs of elements of X are similar to the corresponding distances between pairs of elements of Y.

$$d_{ij}^x \approx d_{ij}^y \tag{4.44}$$

$i, j = 1, \ldots, n$. Just like MDS, Sammon mapping may also be used to produce a feature representation Y from a distance matrix D^x. Y is found by minimizing the error between

D^x and D^y. Candidates for this error functional are

$$E_1 = \frac{1}{\sum\limits_{i=1}^{n} \sum\limits_{j=i+1}^{n} \left(d_{ij}^x\right)^2} \sum\limits_{i=1}^{n} \sum\limits_{j=i+1}^{n} \left(d_{ij}^y - d_{ij}^x\right)^2 \tag{4.45}$$

$$E_2 = \sum\limits_{i=1}^{n} \sum\limits_{j=i+1}^{n} \left(\frac{d_{ij}^y - d_{ij}^x}{d_{ij}^x}\right)^2 \tag{4.46}$$

$$E_3 = \frac{1}{\sum\limits_{i=1}^{n} \sum\limits_{j=i+1}^{n} d_{ij}^x} \sum\limits_{i=1}^{n} \sum\limits_{j=i+1}^{n} \frac{\left(d_{ij}^y - d_{ij}^x\right)^2}{d_{ij}^x} \tag{4.47}$$

The normalization factors in E_1 and E_3 depend only on X and may therefore be ignored in the minimization. E_1 is the absolute quadratic error, E_2 is the relative quadratic error, and E_3 is a compromise between the absolute and relative quadratic error that usually yields the best results, so we will restrict our attention to E_3 here. No closed form solution has been found for minima of E_3, so E_3 may be iteratively minimized by gradient descent or Newton optimization (see the Appendix for more details). For the computation of the first and second order derivatives of E_3 notice that

$$\frac{\partial d_{ij}^y}{\partial y_k} = \frac{\partial}{\partial y_k} \|y_i - y_j\| = \begin{cases} \frac{y_k - y_j}{d_{kj}^y} & \text{if } i = k \\ 0 & \text{otherwise} \end{cases} \tag{4.48}$$

so we obtain

$$\frac{\partial E_3}{\partial y_k} = \frac{2}{\sum\limits_{i=1}^{n} \sum\limits_{j=i+1}^{n} d_{ij}^x} \sum\limits_{\substack{j=1 \\ j \neq k}}^{n} \left(\frac{1}{d_{kj}^x} - \frac{1}{d_{kj}^y}\right)(y_k - y_j) \tag{4.49}$$

$$\frac{\partial^2 E_3}{\partial y_k^2} = \frac{2}{\sum\limits_{i=1}^{n} \sum\limits_{j=i+1}^{n} d_{ij}^x} \sum\limits_{\substack{j=1 \\ j \neq k}}^{n} \left(\frac{1}{d_{kj}^x} - \frac{1}{d_{kj}^y} - \frac{(y_k - y_j)^2}{(d_{kj}^y)^3}\right) \tag{4.50}$$

Consider again the data set from Fig. 4.5. Figure 4.9 (left) shows the pairwise distances d^x given in (4.40). We initialize $Y = \{1, 2, 3, 4\}$, which corresponds to the bottom row in

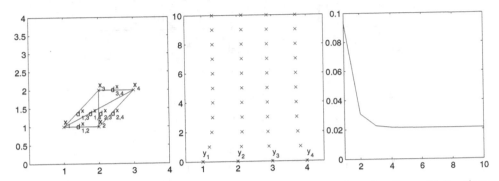

Fig. 4.9 Four points data set: pairwise distances, Sammon mapping after $0, \dots, 10$ iterations, Sammon error function

Fig. 4.9 (center) and yields the distance matrix

$$D^y = \begin{pmatrix} 0\,1\,2\,3 \\ 1\,0\,1\,2 \\ 2\,1\,0\,1 \\ 3\,2\,1\,0 \end{pmatrix} \tag{4.51}$$

and the initial Sammon error

$$E_3 = \frac{1}{3 \cdot 1 + 2 \cdot \sqrt{2} + \sqrt{5}} \cdot \left(2 \cdot \frac{\left(2 - \sqrt{2}\right)^2}{\sqrt{2}} + \frac{\left(3 - \sqrt{5}\right)^2}{\sqrt{5}} \right) \approx 0.0925 \tag{4.52}$$

This is the first (leftmost) value of the Sammon error function shown in Fig. 4.9 (right). For this initialization, the error gradients are

$$\frac{\partial E_3}{\partial y_1} = \frac{2}{3 \cdot 1 + 2 \cdot \sqrt{2} + \sqrt{5}} \cdot \left(-\frac{2 - \sqrt{2}}{\sqrt{2}} - \frac{3 - \sqrt{5}}{\sqrt{5}} \right) \approx -0.1875 \tag{4.53}$$

$$\frac{\partial E_3}{\partial y_2} = \frac{2}{3 \cdot 1 + 2 \cdot \sqrt{2} + \sqrt{5}} \cdot \left(-\frac{2 - \sqrt{2}}{\sqrt{2}} \right) \approx -0.1027 \tag{4.54}$$

$$\frac{\partial E_3}{\partial y_3} = \frac{2}{3 \cdot 1 + 2 \cdot \sqrt{2} + \sqrt{5}} \cdot \left(\frac{2 - \sqrt{2}}{\sqrt{2}} \right) \approx 0.1027 \tag{4.55}$$

$$\frac{\partial E_3}{\partial y_4} = \frac{2}{3 \cdot 1 + 2 \cdot \sqrt{2} + \sqrt{5}} \cdot \left(\frac{3 - \sqrt{5}}{\sqrt{5}} + \frac{2 - \sqrt{2}}{\sqrt{2}} \right) \approx 0.1875 \tag{4.56}$$

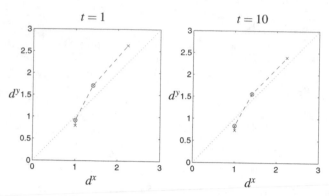

Fig. 4.10 Shepard diagrams for Sammon projection (four points data set)

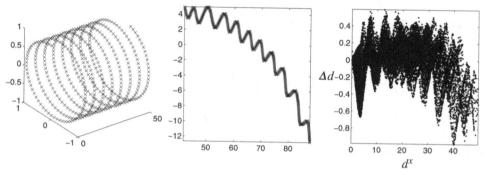

Fig. 4.11 Helix data set, Sammon projection, and projection errors

With step length $\alpha = 1$, gradient descent produces the next estimate $Y \approx$ (1.1875, 2.1027, 2.8973, 3.8125), which corresponds to the four points in the second row in Fig. 4.9 (center). The center and right view of Fig. 4.9 show the values of Y and E_3 for the first ten gradient descent steps. After ten steps we obtain $Y \approx$ (1.3058, 2.1359, 2.8641, 3.6942) and $E_3 \approx 0.0212$. Figure 4.10 shows the Shepard diagrams for the Sammon projection after one and ten gradient descent steps. In contrast to PCA/MDS (Fig. 4.6), the Sammon mapping yields a Shepard diagram where all points are close to the main diagonal but none of them is very close. Figures 4.11 and 4.12 show the Sammon mapping results for the helix (4.42) and bent square (4.43) data sets (Newton's method, random initialization, 100 steps). Compared with PCA/MDS (Figs. 4.7 and 4.8), the Sammon mapping yields lower and unbiased projection errors. The computational effort for Sammon mapping, however, is usually substantially higher than that of PCA/MDS.

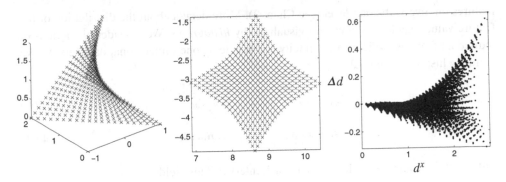

Fig. 4.12 Bent square data set, Sammon projection, and projection errors

4.5 Auto-encoder

The idea of an auto-encoder [4] is to find two functions $f : \mathbb{R}^p \to \mathbb{R}^q$ and $g : \mathbb{R}^q \to \mathbb{R}^p$ that map X to Y and back, so

$$y_k = f(x_k) \tag{4.57}$$

$$x_k \approx g(y_k) \tag{4.58}$$

$k = 1, \ldots, n$. Given the data set X, the parameters of the functions f and g can be found by regression (see Sect. 4.6), where each vector x_k serves as input and output training vector at the same time, hence the name *auto-encoder*.

$$x_k \approx g \circ f(x_k) = g(f(x_k)) \tag{4.59}$$

Once the regression functions f and g are trained, the *forward* transformation function f is used to generate the projected data Y from X using (4.57), and the *backward* transformation function is ignored. Popular regression methods used for auto-encoders include *multi layer perceptrons* (MLP, see Chap. 6). In general, the quality of the mapping using auto-encoders strongly depends on the underlying regression function. Auto-encoders based on neural networks are widely used as layers of *deep learning* architectures [1].

4.6 Histograms

The previous sections presented methods to visualize individual (but possibly high-dimensional) feature vectors. This section deals with visualizing the statistical distribution of features. Statistical measures like the mode, median or mean can be used to describe features on a very abstract level (see Chap. 2). More details about the distribution of the feature values can be obtained and visualized by *histograms*. We consider histograms of individual features, which for simplicity we denote as one-dimensional data sets X. We define m histogram intervals

$$[\xi_1, \xi_2], [\xi_2, \xi_3], \ldots, [\xi_m, \xi_{m+1}] \tag{4.60}$$

$$\xi_1 = \min X, \quad \xi_{m+1} = \max X \tag{4.61}$$

and count the number of values in X in each interval. This yields

$$h_k(X) = |\{\xi \in X \mid \xi_k \leq \xi < \xi_{k+1}\}|, \ k = 1, \ldots, m - 1 \tag{4.62}$$

$$h_m(X) = |\{\xi \in X \mid \xi_m \leq \xi \leq \xi_{m+1}\}| \tag{4.63}$$

where the sum of the counts is the number of data points

$$\sum_{k=1}^{m} h_k(X) = |X| = n \tag{4.64}$$

Equally spaced intervals $\Delta x = (\max X - \min X)/m$ are often used, which yield the interval borders $\xi_k = \min X + (k - 1) \cdot \Delta x, \ k = 1, \ldots, m + 1$. Each count h_k, $k = 1, \ldots, m$ can be visualized as a vertical bar. The left and right borders of each bar represent the lower and upper limits of the corresponding data interval. The height of each bar represents the interval count. A diagram of such bars is called a *histogram*. Figure 4.13 shows the histograms ($m = 100$) of the three features of the data set from Fig. 4.3. The first two features are apparently approximately equally distributed. The third feature contains many values close to zero and to one, which correspond to the upper and lower planes in the scatter plot in Fig. 4.2. Such data accumulations can be easily seen in

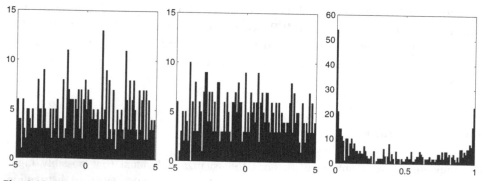

Fig. 4.13 Histograms of the features x, y, z

histograms. Histograms can also be used to estimate the underlying statistical distributions (for example uniform, Gaussian, or Poisson distributions). The parameter m, the number of histogram bins, has to be chosen carefully. If m is too low, then the shape of the distribution cannot be recognized from the histogram. The same observation holds if m is too high, because then most bins are empty or contain only few points. A good choice of m generally depends on the data. Some rules of thumb often yield good estimates for m:

- Sturgess [11] (based on the number of data)

$$m = 1 + \log_2 n \qquad (4.65)$$

- Scott [10] (based on the standard deviation, tailored for Gaussian distributions)

$$m = \frac{3.49 \cdot s}{\sqrt[3]{n}} \qquad (4.66)$$

- Freedman and Diaconis [3] (based on the middle 50% quantile, also suitable for long tailed distributions)

$$m = \frac{2 \cdot (Q_{75\%} - Q_{25\%})}{\sqrt[3]{n}} \qquad (4.67)$$

with quantiles defined by

$$|\{x \in X \mid x \leq Q_\varphi\}| = \varphi \cdot n \qquad (4.68)$$

Instead of equally spaced bins, the histogram intervals may have different widths, for example to obtain a finer resolution in areas with high data densities [5]. Histogram intervals may be defined as $1/m$ quantiles, so that each bin contains the same number of data. In the visualization of such non-equally spaced histograms the height of each bar does not correspond to the number of data but to the number of data divided by the bin width, so the area of each bar reflects the corresponding number of data.

Every datum between the lower and upper bounds of a histogram interval is counted for the respective bin, whether it is close to the interval center or close to the border. A *fuzzy histogram* [8] partially counts data for several neighboring bins. For example, a datum at the border between two bins may be counted as half for one and half for the other bin. More generally, fuzzy histograms use data counts in *fuzzy intervals* defined by membership functions $\mu : X \to [0, 1]$. For example, triangular membership functions are

Fig. 4.14 Triangular
membership functions

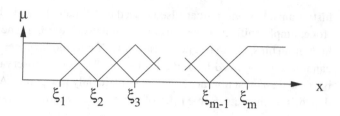

often used (Fig. 4.14):

$$\mu_1(x) = \begin{cases} 1 & \text{if } x < \xi_1 \\ \frac{\xi_2 - x}{\xi_2 - \xi_1} & \text{if } \xi_1 \leq x < \xi_2 \\ 0 & \text{if } x \geq \xi_2 \end{cases} \tag{4.69}$$

$$\mu_k(x) = \begin{cases} 0 & \text{if } x < \xi_{k-1} \\ \frac{x - \xi_{k-1}}{\xi_k - \xi_{k-1}} & \text{if } \xi_{k-1} \leq x < \xi_k \\ \frac{\xi_{k+1} - x}{\xi_{k+1} - \xi_k} & \text{if } \xi_k \leq x < \xi_{k+1} \\ 0 & \text{if } x \geq \xi_{k+1} \end{cases} \tag{4.70}$$

$$k = 2, \ldots, m - 1$$

$$\mu_m(x) = \begin{cases} 0 & \text{if } x < \xi_{m-1} \\ \frac{x - \xi_{m-1}}{\xi_m - \xi_{m-1}} & \text{if } \xi_{m-1} \leq x < \xi_m \\ 1 & \text{if } x \geq \xi_m \end{cases} \tag{4.71}$$

Such membership functions assign each datum x to a certain extent to one or two intervals, where the sum of counts is one. In general, the counts of fuzzy histograms are computed as

$$\tilde{h}_k(X) = \sum_{x \in X} \mu_k(x) \tag{4.72}$$

Conventional histograms are special cases of fuzzy histograms for complementary rectangular membership functions.

4.7   Spectral Analysis

The purpose of data visualization is to emphasize important data characteristics. Important characteristics of *time series* data are spectral features such as the amplitude and phase spectra. These spectra are motivated by the Fourier theorem: Every continuously differentiable function f can be decomposed into cosine and sine components according to

$$f(x) = \int_0^\infty (a(y) \cos xy + b(y) \sin xy) \, dy \quad \text{with} \tag{4.73}$$

$$a(y) = \frac{1}{\pi} \int_{-\infty}^\infty f(u) \cos yu \, du \tag{4.74}$$

$$b(y) = \frac{1}{\pi} \int_{-\infty}^\infty f(u) \sin yu \, du \tag{4.75}$$

Based on this theorem the *Fourier cosine transform* $F_c(y)$ and the *Fourier sine transform* $F_s(y)$ are defined as

$$F_c(y) = \sqrt{\frac{2}{\pi}} \int_0^\infty f(x) \cos xy \, dx \tag{4.76}$$

$$f(x) = \sqrt{\frac{2}{\pi}} \int_0^\infty F_c(y) \cos xy \, dy \tag{4.77}$$

$$F_s(y) = \sqrt{\frac{2}{\pi}} \int_0^\infty f(x) \sin xy \, dx \tag{4.78}$$

$$f(x) = \sqrt{\frac{2}{\pi}} \int_0^\infty F_s(y) \sin xy \, dy \tag{4.79}$$

The Fourier cosine and sine transforms can be applied to discrete functions by substituting $x = k \cdot T$ and $y = l \cdot \omega$ in ((4.76)–(4.79)), where T is a time constant and ω is a frequency constant. The data x_k, $k = 1, \ldots, n$, are considered to represent equidistant samples of the continuous function f, so $f(k \cdot T) = x_k$, $k = 1, \ldots, n$. The Fourier transform then yields the Fourier cosine and sine transform data $F_c(l \cdot \omega) = y_l^c$ and $F_s(l \cdot \omega) = y_l^s$, $l = 1, \ldots, m$.

$$y_l^c = \frac{2}{n} \sum_{k=1}^n x_k \cos kl\omega T \tag{4.80}$$

$$x_k' = \frac{n\omega T}{\pi} \sum_{l=1}^m y_l^c \cos kl\omega T \tag{4.81}$$

$$y_l^s = \frac{2}{n} \sum_{k=1}^{n} x_k \sin kl\omega T \qquad (4.82)$$

$$x_k' = \frac{n\omega T}{\pi} \sum_{l=1}^{m} y_l^s \sin kl\omega T \qquad (4.83)$$

Figure 4.15 (left) shows a graph of the $n = 5000$ points in the artificial data set

$$X = \{(i, \cos(0.001 \cdot i) + 0.5 \cdot \cos(0.01 \cdot i - 1)) \mid i \in \{1, \ldots, 5000\}\} \qquad (4.84)$$

that comprises two periodic sequences, one with lower frequency and higher amplitude and the other with higher frequency and lower amplitude. Imagine that such a data set might be generated from sales figures with weekly and seasonal fluctuations. The center column in Fig. 4.15 shows the absolute values of the Fourier cosine transform (4.80) (top) and the Fourier sine transform (4.82) (bottom) for $m = 100$ and $\omega T = 10^{-4}$. The parameters ω and T always occur as products ωT, so ωT can be considered as only one parameter. Here, $\omega T = 10^{-4}$. The local maxima of both transforms are at about y_{10} and y_{100}, which corresponds to the frequencies $10 \cdot \omega T = 0.001$ and $100 \cdot \omega T = 0.01$ and thus to the frequencies in the cosine terms of (4.84). The values of the maxima are $y_9^c \approx 0.9722$, $y_{98}^c \approx$

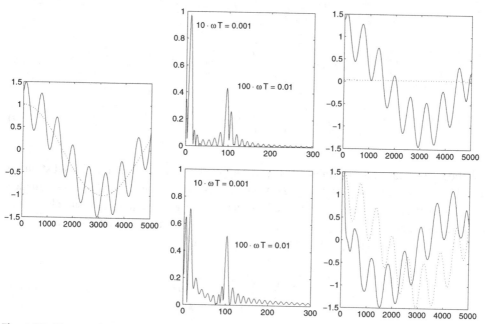

Fig. 4.15 Time series data, absolute Fourier transforms, and reverse transforms (top: cosine, bottom: sine)

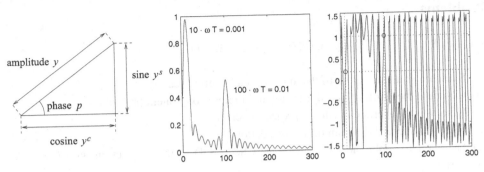

Fig. 4.16 Orthogonal decomposition, amplitude and phase spectrum

0.4313, $y_{14}^s \approx 0.7075$, and $y_{103}^s \approx 0.5104$, which roughly correspond to the amplitudes of the cosine terms at (4.84), but especially y_{98}^c and y_{14}^s do not match well with the amplitudes 0.5 and 1. The right column in Fig. 4.15 shows the reverse Fourier cosine transform (4.81) (top) and the reverse Fourier sine transform (4.83) (bottom). The reverse Fourier cosine transform approximately matches the original time series. The error (displayed as a dashed curve) is almost zero. However, the reverse Fourier sine transform is very different from the original time series (displayed here as the dashed curve).

The Fourier cosine transform y^c and the Fourier sine transform y^s can be viewed as the legs of a right-angled triangle, as shown in Fig. 4.16 (left). The *Fourier amplitude spectrum* y is defined as the hypotenuse and the *Fourier phase spectrum* is defined as the angle opposed to the sine of this triangle.

$$y_l = \sqrt{(y_l^c)^2 + (y_l^s)^2} \tag{4.85}$$

$$p_l = \arctan \frac{y_l^s}{y_l^c} \tag{4.86}$$

Figure 4.16 (center) shows the Fourier amplitude spectrum (4.85) of our example time series. The local maxima are at $y_{10} \approx 0.9784$ and $y_{101} \approx 0.5261$, and match quite well the frequencies and amplitudes of the cosine terms in (4.84). Figure 4.16 (right) shows the Fourier phase spectrum (4.86). The angles at the amplitude maxima are $p_{10} \approx 0.2073$ and $p_{100} \approx 1.0320$, and approximately correspond with the original angle offsets of the cosine terms in (4.84). Thus, given time series data, Fourier analysis allows us to compute the amplitude spectrum $Y = \{y_1, \ldots, y_m\} \subset \mathbb{R}$ and the phase spectrum $P = \{p_1, \ldots, p_m\} \subset \mathbb{R}$, that represent the frequencies, amplitudes, and phase angles of the spectral components of the time series.

Problems

4.1 Let $X = \{(-2, 1), (-1, 2), (0, 0), (1, -2), (2, -1)\}$.

(a) Sketch a scatter diagram of this data set and its eigenvectors.
(b) Compute the result of one-dimensional principal component projection.
(c) Compute the average quadratic projection error.
(d) Sketch a Shepard diagram of this projection.
(e) We want to apply the kernel trick (see Sect. 8.4) to this projection with Gaussian kernels

$$k(x, y) = e^{-\frac{\|x-y\|^2}{\sigma^2}}, \quad \sigma = 1$$

Which operation in the projection is replaced by the kernel function?
(f) Give the equation for this kernelized projection.
(g) What are the results of this kernelized projection for the data set X?
(h) How do these results indicate that this kernelized projection is a *nonlinear* mapping?

4.2 Consider Sammon mapping of a dissimilarity matrix D^X.

(a) For which values of q can Sammon mapping yield a q-dimensional representation of $Y \subset \mathbb{R}^q$ with zero error for Euclidean distances for *any* 4×4 dissimilarity matrix D^X?
(b) Sketch a Shepard diagram for such a mapping.
(c) Explain why this does not work for $D^X = \begin{pmatrix} 0\ 1\ 1\ 1 \\ 1\ 0\ 1\ 1 \\ 1\ 1\ 0\ 3 \\ 1\ 1\ 3\ 0 \end{pmatrix}$.

4.3 Consider an auto-encoder $X \to Y \to X'$, where $X, X' \in \mathbb{R}^2$, $Y \in \mathbb{R}$, with

$$y = f(x) = tanh\left(\frac{x^{(1)} + x^{(2)}}{2}\right).$$

(a) Find a suitable function $x' = g(y)$.
(b) Calculate the average quadratic error of the transformation $g \circ f$ for the data set $X = \{(0, 0), (0, 1), (1, 0), (1, 1)\}$.
(c) Which other projection method would for this data set X yield the same X'?

References

1. Y. Bengio. Learning deep architectures for AI. *Foundations and Trends in Machine Learning*, 2(1):1–127, 2009.
2. R. O. Duda and P. E. Hart. *Pattern Classification and Scene Analysis*. Wiley, New York, 1973.
3. D. Freedman and P. Diaconis. On the histogram as a density estimator: L2 theory. *Probability Theory and Related Fields*, 57(4):453–476, December 1981.
4. G. E. Hinton. Connectionist learning procedures. *Artificial Intelligence*, 40:185–234, 1989.
5. A. K. Jain. *Fundamentals of Digital Image Processing*. Prentice Hall, Englewood Cliffs, 1986.
6. D. Keim, G. Andrienko, J.-D. Fekete, C. Görg, J. Kohlhammer, and G. Melançon. Visual analytics: Definition, process, and challenges. In *Information visualization*, pages 154–175. Springer, 2008.
7. K. Pearson. On lines and planes of closest fit to systems of points in space. *Philosophical Magazine*, 2(6):559–572, 1901.
8. T. A. Runkler. Fuzzy histograms and fuzzy chi–squared tests for independence. In *IEEE International Conference on Fuzzy Systems*, volume 3, pages 1361–1366, Budapest, July 2004.
9. J. W. Sammon. A nonlinear mapping for data structure analysis. *IEEE Transactions on Computers*, C-18(5):401–409, 1969.
10. D. W. Scott. On optimal and data–based histograms. *Biometrika*, 66(3):605–610, 1979.
11. H. A. Sturges. The choice of a class interval. *Journal of the American Statistical Association*, pages 65–66, 1926.
12. W. S. Torgerson. *Theory and Methods of Scaling*. Wiley, New York, 1958.
13. J. W. Tukey. *Exploratory Data Analysis*. Addison Wesley, Reading, 1987.
14. G. Young and A. S. Householder. Discussion of a set of points in terms of their mutual distances. *Psychometrika*, 3:19–22, 1938.

Correlation

Abstract

Correlation quantifies the relationship between features. Linear correlation methods are robust and computationally efficient but detect only linear dependencies. Nonlinear correlation methods are able to detect nonlinear dependencies but need to be carefully parametrized. As a popular example for nonlinear correlation we present the chi-square test for independence that can be applied to continuous features using histogram counts. Nonlinear correlation can also be quantified by the cross-validation error of regression models. Correlation does not imply causality. Spurious correlations may lead to wrong conclusions. If the underlying features are known, then spurious correlations may be compensated by partial correlation methods.

5.1   Linear Correlation

Correlation quantifies the relationship between features. The purpose of correlation analysis is to understand the dependencies between features, so that observed effects can be explained or desired effects can be achieved. For example, for a production plant correlation analysis may yield the features that correlate with the product quality, so the target quality can be achieved by systematically modifying the most relevant features. Figure 5.1 gives an overview of the correlation measures presented in this chapter. First, we focus on the linear correlation between pairs of features. In (2.20) and (4.8) we introduced the covariance matrix C of a data set $X \subset \mathbb{R}^p$, where each matrix element c_{ij} denotes the covariance between the features $x^{(i)}$ and $x^{(j)}$, $i, j = 1, \ldots, p$.

$$c_{ij} = \frac{1}{n-1} \sum_{k=1}^{n} (x_k^{(i)} - \bar{x}^{(i)})(x_k^{(j)} - \bar{x}^{(j)}) \tag{5.1}$$

© Springer Fachmedien Wiesbaden GmbH, part of Springer Nature 2020
T. A. Runkler, *Data Analytics*, https://doi.org/10.1007/978-3-658-29779-4_5

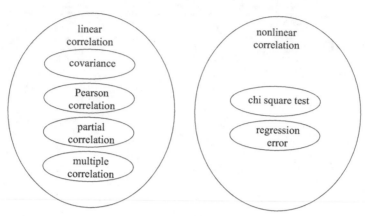

Fig. 5.1 Some important correlation measures

If c_{ij} is positive large, then there is a strong positive dependency between $x^{(i)}$ and $x^{(j)}$, i.e. high values of $x^{(i)}$ coincide with high values of $x^{(j)}$, and low values of $x^{(i)}$ coincide with low values of $x^{(j)}$. If c_{ij} is negative large, then there is a strong negative dependency, i.e. high values of $x^{(i)}$ coincide with low values of $x^{(j)}$ and vice versa. If c_{ij} is close to zero, then there is a weak dependency between $x^{(i)}$ and $x^{(j)}$. If a feature is multiplied by a constant factor α, for example when it is given in different unit (e.g. meters instead of kilometers), then the covariance between this feature and any other feature will also increase by a factor α, although we do not expect this feature to make more useful contributions to data analysis. The *Pearson correlation coefficient* compensates the effect of constant scaling by dividing the covariance by the product of the standard deviations of both features.

$$s_{ij} = \frac{c_{ij}}{s^{(i)}s^{(j)}} \tag{5.2}$$

$$= \frac{\sum_{k=1}^{n}(x_k^{(i)} - \bar{x}^{(i)})(x_k^{(j)} - \bar{x}^{(j)})}{\sqrt{\left(\sum_{k=1}^{n}(x_k^{(i)} - \bar{x}^{(i)})^2\right)\left(\sum_{k=1}^{n}(x_k^{(j)} - \bar{x}^{(j)})^2\right)}} \tag{5.3}$$

$$= \frac{\sum_{k=1}^{n}x_k^{(i)}x_k^{(j)} - n\,\bar{x}^{(i)}\bar{x}^{(j)}}{\sqrt{\left(\sum_{k=1}^{n}\left(x_k^{(i)}\right)^2 - n\left(\bar{x}^{(i)}\right)^2\right)\left(\sum_{k=1}^{n}\left(x_k^{(j)}\right)^2 - n\left(\bar{x}^{(j)}\right)^2\right)}} \tag{5.4}$$

The standard deviations are the square roots of the variances, i.e. the square roots of the diagonal elements of the covariance matrix, $s^{(i)} = \sqrt{c_{ii}}$, so the (Pearson) correlation matter can be directly computed from the covariance matrix.

$$s_{ij} = \frac{c_{ij}}{\sqrt{c_{ii}c_{jj}}} \tag{5.5}$$

where $s_{ij} \in [-1, 1]$. If $s_{ij} \approx 1$ then there is a strong positive correlation between $x^{(i)}$ and $x^{(j)}$. If $s_{ij} \approx -1$ then there is a strong negative correlation. If $s_{ij} \approx 0$ then $x^{(i)}$ and $x^{(j)}$ are (almost) independent, so correlation can be interpreted as the opposite of independence. Notice that for μ-σ-standardized data, covariance and correlation are equal, since

$$c_{ii} = c_{jj} = 1 \quad \Rightarrow \quad s_{ij} = \frac{c_{ij}}{\sqrt{c_{ii}c_{jj}}} = c_{ij} \tag{5.6}$$

5.2 Correlation and Causality

Correlation does not imply causality. A correlation between x and y may indicate four different causal scenarios or a combination of these:

1. coincidence
2. x causes y
3. y causes x
4. z causes both x and y

Even though the data suggest a correlation, this might be a coincidence, so x and y do not possess any causal connection (scenario 1). If there is a causal connection between x and y, then correlation does not distinguish whether x causes y or y causes x (scenarios 2 and 3), for example a correlation between the consumption of diet drinks and obesity does not tell us whether diet drinks cause obesity or obesity causes people to drink diet drinks. Finally, there might be no direct causal connection between x and y: instead both x and y may be caused by a third variable z (scenario 4). For example a correlation between forest fires and harvest does not imply that forest fires increase harvest nor that harvest causes forest fires. Instead, it may be that both forest fires and harvest are caused by sunny weather. This is called *spurious correlation* [4] or *third cause fallacy*. Correlation analysis does not distinguish between these scenarios is valid, so often additional expert knowledge is required.

If $x^{(i)}$ and $x^{(j)}$ are correlated and also both correlated with $x^{(k)}$ (like in the spurious correlation scenario), then we might want to know the correlation between $x^{(i)}$ and $x^{(j)}$ *without* the influence of $x^{(k)}$. This is called the *partial* or *conditional correlation* which is defined as

$$s_{ij|k} = \frac{s_{ij} - s_{ik}s_{jk}}{\sqrt{(1 - s_{ik}^2)(1 - s_{jk}^2)}} \tag{5.7}$$

for $s_{ik}, s_{jk} \neq \pm 1$. The correlation between $x^{(i)}$ and $x^{(j)}$ without the influence of the two features $x^{(k)}$ and $x^{(l)}$ is called *bipartial correlation* defined as

$$s_{i|k,\,j|l} = \frac{s_{ij} - s_{ik}s_{jk} - s_{il}s_{jl} + s_{ik}s_{kl}s_{jl}}{\sqrt{(1 - s_{ik}^2)(1 - s_{jl}^2)}} \tag{5.8}$$

for $s_{ik}, s_{jl} \neq \pm 1$. The correlation of $x^{(i)}$ with a whole group of features $x^{(j_1)}, \ldots, x^{(j_q)}$ is called *multiple correlation* and is defined as

$$s_{i,(j_1,\ldots,j_q)} = \sqrt{(s_{ij_1} \ldots s_{ij_q}) \cdot \begin{pmatrix} 1 & s_{j_2 j_1} & \cdots & s_{j_1 j_q} \\ s_{j_1 j_2} & 1 & \cdots & s_{j_2 j_q} \\ \vdots & \vdots & \ddots & \vdots \\ s_{j_1 j_q} & s_{j_2 j_q} & \cdots & 1 \end{pmatrix}^{-1} \begin{pmatrix} s_{ij_1} \\ s_{ij_2} \\ \vdots \\ s_{ij_q} \end{pmatrix}} \tag{5.9}$$

For $q = 1$, multiple correlation becomes the simple correlation.

$$s_{i,(j_1)} = |s_{ij_1}| \tag{5.10}$$

For $q = 2$ we obtain

$$s_{i,(j_1,j_2)} = \sqrt{\frac{s_{ij_1}^2 + s_{ij_2}^2 - 2s_{ij_1}s_{ij_2}s_{j_1 j_2}}{1 - s_{j_1 j_2}^2}} \tag{5.11}$$

for $s_{j_1 j_2} \neq \pm 1$. Notice that $s_{ij} = [-1, 1]$ holds for the (simple) correlation, but not necessarily for the partial, bipartial, and multiple correlations. For more information about linear correlation analysis we refer to statistics textbooks, for example [1].

5.3 Chi-Square Test for Independence

The correlation methods presented in the previous section assume linear dependency between features. Even for pairs of features with a very strong nonlinear dependency, the linear correlation may be small or even zero. A method to quantify the nonlinear correlation between features is the chi-square test for independence [2]. To quantify the nonlinear correlation between two continuous features $x^{(1)}$ and $x^{(2)}$, we first compute the histograms of $x^{(1)}$ and $x^{(2)}$ for r and s bins, respectively.

$$h^{(1)} = (h_1^{(1)}, \ldots, h_r^{(1)}), \quad h^{(2)} = (h_1^{(2)}, \ldots, h_s^{(2)}) \tag{5.12}$$

Fig. 5.2 Counts for the
chi-square test for
independence

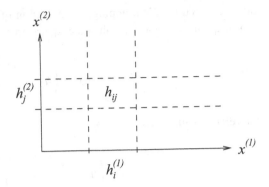

Then we count the number of data that fall into each combination of the ith bin of $x^{(1)}$, $i = 1, \ldots, r$, and the jth bin of $x^{(2)}$, $j = 1, \ldots, s$, denote this count as h_{ij}, and write these counts as a matrix

$$H = \begin{pmatrix} h_{11} & h_{12} & \cdots & h_{1s} \\ h_{21} & h_{22} & \cdots & h_{2s} \\ \vdots & \vdots & \ddots & \vdots \\ h_{r1} & h_{r2} & \cdots & h_{rs} \end{pmatrix} \tag{5.13}$$

Figure 5.2 illustrates the counts $h_i^{(1)}$, $h_j^{(2)}$, and h_{ij}, $i = 1, \ldots, r$, $j = 1, \ldots, s$. The histograms of $x^{(1)}$ and $x^{(2)}$ are the row and column sums of H, respectively.

$$\sum_{j=1}^{s} h_{ij} = h_i^{(1)}, \quad i = 1, \ldots, r \tag{5.14}$$

$$\sum_{i=1}^{r} h_{ij} = h_j^{(2)}, \quad j = 1, \ldots, s \tag{5.15}$$

If the features are independent, then the probability of a data point falling into the bin combination h_{ij} is equal to the product of the probability of falling into bin $h_i^{(1)}$ and the probability of falling into bin $h_j^{(2)}$, so

$$\frac{h_{ij}}{n} = \frac{h_i^{(1)}}{n} \cdot \frac{h_j^{(2)}}{n} \quad \Rightarrow \quad h_{ij} = \frac{h_i^{(1)} \cdot h_j^{(2)}}{n} \tag{5.16}$$

where

$$n = \sum_{i=1}^{r} \sum_{j=1}^{s} h_{ij} = \sum_{i=1}^{r} h_i^{(1)} = \sum_{j=1}^{s} h_j^{(2)} \tag{5.17}$$

Similar to Sammon's mapping, the deviation of h_{ij} from complete independence (5.16) can be quantified using the absolute square error

$$E_1 = \left(h_{ij} - \frac{h_i^{(1)} \cdot h_j^{(2)}}{n} \right)^2 \tag{5.18}$$

the relative square error

$$E_2 = \left(h_{ij} - \frac{h_i^{(1)} \cdot h_j^{(2)}}{n} \right)^2 \bigg/ \left(\frac{h_i^{(1)} \cdot h_j^{(2)}}{n} \right)^2 \tag{5.19}$$

or a compromise between absolute and relative square error

$$E_3 = \left(h_{ij} - \frac{h_i^{(1)} \cdot h_j^{(2)}}{n} \right)^2 \bigg/ \left(\frac{h_i^{(1)} \cdot h_j^{(2)}}{n} \right) \tag{5.20}$$

Just as in Sammon's mapping we choose the compromise E_3 and obtain the chi-square test statistic

$$\chi^2 = \frac{1}{n} \sum_{i=1}^{r} \sum_{j=1}^{s} \frac{\left(n \cdot h_{ij} - h_i^{(1)} \cdot h_j^{(2)} \right)^2}{h_i^{(1)} \cdot h_j^{(2)}} \tag{5.21}$$

The hypothesis that the features are independent is rejected if

$$\chi^2 > \chi^2(1 - \alpha, r - 1, s - 1) \tag{5.22}$$

where α is the significance level. Values of this distribution function are tabulated and available in [1], for example. Small values of χ^2 confirm the hypothesis that the features are independent. The chi-square distribution is strictly monotonically increasing, so the lower χ^2, the lower the probability for stochastic independence, and the higher the degree of stochastic independence between the considered pair of features. Therefore, to produce a list of pairs of features in order of their nonlinear correlation, it is not necessary to specify the significance level α and to compute $\chi^2(1 - \alpha, r - 1, s - 1)$, but it is sufficient to sort the plain χ^2 values for the feature pairs in descending order.

The number of bins has to be chosen carefully when using the chi-square test for independence, just like for histograms. Nonlinear dependencies cannot be recognized if there are either too few or too many bins. If there are too few bins, then the resulting coarse grid cannot adequately represent the functional dependency. If there are too many bins, then most bins will be empty or contain only a few points. A good choice for the number of bins is often obtained using the rules of thumb given in the histogram section.

Also *fuzzy* bins may be used in the chi-square test for independence, which yields the *fuzzy chi-square test for independence* [3].

As an alternative to the chi-square test for independence, nonlinear correlation can also be quantified by building regression models and computing the cross-validation error. This approach is discussed in more detail in the next chapter.

Problems

5.1 Pearson correlation analysis yields to the following findings. Explain these results and suggest ways to improve the analysis.

(a) Ice cream sales are highly correlated with accidents in swimming pools.
(b) Average flight delays are highly correlated with the winning numbers in the lottery drawing.
(c) The fuel consumption of a car is only weakly correlated with the speed.

5.2 Six flowers receive daily water amounts between 0 and 100%. After 1 month they have grown according to the following table:

Flowers number	1	2	3	4	5	6
Daily water amount	10%	40%	40%	60%	60%	90%
Growth	10%	60%	70%	60%	70%	10%

(a) What is the Pearson correlation between daily water amount and growth.
(b) Would the chi-square test for independence with two bins for daily water amount (0,50,100%) and three bins for growth (0,33,66,100%) indicate a high or low correlation?
(c) Would the chi-square test for independence with three bins for daily water amount (0,33,66,100%) and two bins for growth (0,50,100%) indicate a high or low correlation?
(d) How do you interpret the three results?

5.3 The chi-square formula for two variables is

$$\chi^2 = \frac{1}{n} \sum_{i=1}^{r} \sum_{j=1}^{s} \frac{\left(n \cdot h_{ij} - h_i^{(1)} \cdot h_j^{(2)}\right)^2}{h_i^{(1)} \cdot h_j^{(2)}}$$

What would be the formula if we consider three variables?

References

1. D. Freedman, R. Pisani, and R. Purves. *Statistics*. W. W. Norton & Company, New York, 2007.
2. K. Pearson. On the criterion that a given system of deviations from the probable in the case of a correlated system of variables is such that it can be reasonably supposed to have arisen from random sampling. *Philosophical Magazine*, 50(302):157–175, 1900.
3. T. A. Runkler. Fuzzy histograms and fuzzy chi–squared tests for independence. In *IEEE International Conference on Fuzzy Systems*, volume 3, pages 1361–1366, Budapest, July 2004.
4. H. A. Simon. Spurious correlation: A causal interpretation. *Journal of the American Statistical Association*, 49:467–479, 1954.

Regression

6

Abstract

Regression estimates functional dependencies between features. Linear regression models can be efficiently computed from covariances but are restricted to linear dependencies. Substitution allows to identify specific types of nonlinear dependencies by linear regression. Robust regression finds models that are robust against inliers or outliers. A popular class of nonlinear regression methods are universal approximators. We present two well-known examples for universal approximators from the field of artificial neural networks: the multilayer perceptron and radial basis function networks. Universal approximators can realize arbitrarily small training errors, but cross-validation is required to find models with low validation errors that generalize well on other data sets. Feature selection allows us to include only relevant features in regression models leading to more accurate models.

6.1 Linear Regression

The correlation methods discussed in the previous chapter quantify the degree of relationship between features. In contrast to correlation, regression estimates the actual functional dependency between features. For example, if correlation analysis has identified the features that mostly influence product quality, then regression models indicate to which specific values the features should be set to achieve a given target quality. Figure 6.1 gives an overview of the regression methods presented in this chapter. Just as in the previous chapter we first focus on *linear* methods. *Linear regression* identifies a linear functional dependency between features. The linear approximation of a function of *one* feature $x^{(i)} = f(x^{(j)})$, $i, j \in \{1, \ldots, p\}$, can be written as

$$x_k^{(i)} \approx a \cdot x_k^{(j)} + b \tag{6.1}$$

© Springer Fachmedien Wiesbaden GmbH, part of Springer Nature 2020
T. A. Runkler, *Data Analytics*, https://doi.org/10.1007/978-3-658-29779-4_6

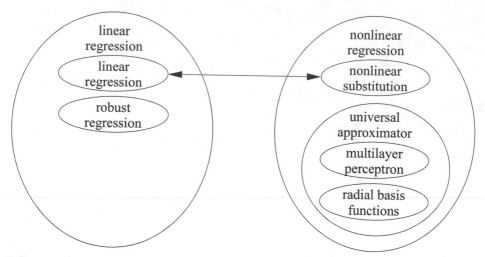

Fig. 6.1 Some important regression methods

so linear regression has to estimate the parameters $a, b \in \mathbb{R}$ from X by minimizing a suitable error functional. Conventional linear regression uses the average quadratic regression error

$$E = \frac{1}{n} \sum_{k=1}^{n} e_k^2 = \frac{1}{n} \sum_{k=1}^{n} \left(x_k^{(i)} - a \cdot x_k^{(j)} - b \right)^2 \tag{6.2}$$

One necessary criterion for local extrema of E is

$$\frac{\partial E}{\partial b} = -\frac{2}{n} \sum_{k=1}^{n} \left(x_k^{(i)} - a \cdot x_k^{(j)} - b \right) = 0 \quad \Rightarrow \quad b = \bar{x}^{(i)} - a \cdot \bar{x}_k^{(j)} \tag{6.3}$$

So we can write the regression error as

$$E = \frac{1}{n} \sum_{k=1}^{n} \left(x_k^{(i)} - \bar{x}^{(i)} - a(x_k^{(j)} - \bar{x}^{(j)}) \right)^2 \tag{6.4}$$

The other necessary criterion for local extrema of E is

$$\frac{\partial E}{\partial a} = -\frac{2}{n} \sum_{k=1}^{n} (x_k^{(j)} - \bar{x}^{(j)}) \left(x_k^{(i)} - \bar{x}^{(i)} - a(x_k^{(j)} - \bar{x}^{(j)}) \right) = 0 \tag{6.5}$$

which yields

$$a = \frac{\sum_{k=1}^{n}(x_k^{(i)} - \bar{x}^{(i)})(x_k^{(j)} - \bar{x}^{(j)})}{\sum_{k=1}^{n}(x_k^{(j)} - \bar{x}^{(j)})^2} = \frac{c_{ij}}{c_{jj}}$$ (6.6)

So all pairwise linear regression models of a data set X can be immediately computed from the means of the features and the covariance matrix of X.

The linear approximation of a function of $m \in \{2, 3, \ldots\}$ features $x^{(i)} = f(x^{(j_1)}, \ldots, x^{(j_m)})$, $i, j_1, \ldots, j_m \in \{1, \ldots, p\}$, can be written as

$$x_k^{(i)} \approx \sum_{l=1}^{m} a_l \cdot x_k^{(j_l)} + b$$ (6.7)

and the parameters $a_1, \ldots, a_m, b \in \mathbb{R}$ can be found by minimizing

$$E = \frac{1}{n}\sum_{k=1}^{n} e_k^2 = \frac{1}{n}\sum_{k=1}^{n}\left(x_k^{(i)} - \sum_{l=1}^{m} a_l \cdot x_k^{(j_l)} - b\right)^2$$ (6.8)

One necessary criterion for local extrema of E yields

$$\frac{\partial E}{\partial b} = -\frac{2}{n}\sum_{k=1}^{n}\left(x_k^{(i)} - \sum_{l=1}^{m} a_l \cdot x_k^{(j_l)} - b\right) = 0 \quad \Rightarrow \quad b = \bar{x}^{(i)} - \sum_{l=1}^{m} a_l \cdot \bar{x}^{(j_l)}$$ (6.9)

from which

$$E = \frac{1}{n}\sum_{k=1}^{n}\left(x_k^{(i)} - \bar{x}^{(i)} - \sum_{l=1}^{m} a_l \cdot (x_k^{(j_l)} - \bar{x}^{(j_l)})\right)^2$$ (6.10)

The other necessary criterion for local extrema of E is

$$\frac{\partial E}{\partial a_r} = -\frac{2}{n}\sum_{k=1}^{n}(x_k^{(j_r)} - \bar{x}^{(j_r)})\left(x_k^{(i)} - \bar{x}^{(i)} - \sum_{l=1}^{m} a_l \cdot (x_k^{(j_l)} - \bar{x}^{(j_l)})\right) = 0$$ (6.11)

$r = 1, \ldots, m$, which can be written as a system of linear equations

$$\sum_{l=1}^{m} a_l \sum_{k=1}^{n}(x_k^{(j_l)} - \bar{x}^{(j_l)})(x_k^{(j_r)} - \bar{x}^{(j_r)}) = \sum_{k=1}^{n}(x_k^{(i)} - \bar{x}^{(i)})(x_k^{(j_r)} - \bar{x}^{(j_r)})$$ (6.12)

$$\Leftrightarrow \quad \sum_{l=1}^{m} a_l c_{j_l j_r} = c_{i j_r}, \quad r = 1, \dots, m \tag{6.13}$$

which can be solved by various algorithms for solving systems of linear equations such as Gaussian elimination or Cramer's rule. Thus, all the parameters needed for multiple linear regression can be immediately computed from the means of the features and the covariance matrix of X.

An equivalent result for linear regression is obtained by writing the regression problem (6.7) in matrix form. With (6.9) we denote

$$Y = \begin{pmatrix} x_1^{(i)} - \bar{x}^{(i)} \\ \vdots \\ x_k^{(i)} - \bar{x}^{(i)} \end{pmatrix} \quad X = \begin{pmatrix} x_1^{(j_1)} - \bar{x}^{(j_1)} & \dots & x_1^{(j_m)} - \bar{x}^{(j_m)} \\ \vdots & \ddots & \vdots \\ x_k^{(j_1)} - \bar{x}^{(j_1)} & \dots & x_k^{(j_m)} - \bar{x}^{(j_m)} \end{pmatrix} \quad A = \begin{pmatrix} a_1 \\ \vdots \\ a_m \end{pmatrix} \tag{6.14}$$

and write (6.7) as

$$Y = X \cdot A \tag{6.15}$$

$$X^T \cdot Y = X^T \cdot X \cdot A \tag{6.16}$$

$$(X^T \cdot X)^{-1} \cdot X^T \cdot Y = A \tag{6.17}$$

where $(X^T \cdot X)^{-1} \cdot X^T$ is called the *pseudo inverse* of X, so the regression parameters A can be computed as the product of the pseudo inverse of X and the matrix Y.

As an example for multiple linear regression consider the data set

$$X = \begin{pmatrix} 6 & 4 & -2 \\ 2 & 1 & -1 \\ 0 & 0 & 0 \\ 0 & 1 & 1 \\ 2 & 4 & 2 \end{pmatrix} \tag{6.18}$$

We want to estimate a linear function $x^{(1)} = f(x^{(2)}, x^{(3)})$, so

$$x_k^{(1)} \approx \bar{x}^{(1)} + a_1 (x_k^{(2)} - \bar{x}^{(2)}) + a_2 (x_k^{(3)} - \bar{x}^{(3)}) \tag{6.19}$$

The mean values of the features are

$$\bar{x}^{(1)} = \frac{6+2+2}{5} = 2 \tag{6.20}$$

$$\bar{x}^{(2)} = \frac{4+1+1+4}{5} = 2 \tag{6.21}$$

$$\bar{x}^{(3)} = \frac{-2-1+1+2}{5} = 0 \tag{6.22}$$

The covariance matrix of X is

$$C = \begin{pmatrix} 6 & 3.5 & -2.5 \\ 3.5 & 3.5 & 0 \\ -2.5 & 0 & 2.5 \end{pmatrix} \tag{6.23}$$

The linear equations to determine a_1 and a_2 are

$$c_{22}a_1 + c_{32}a_2 = c_{12} \tag{6.24}$$

$$c_{23}a_1 + c_{33}a_2 = c_{13} \tag{6.25}$$

$$(6.24) \quad \Leftrightarrow \quad 3.5\,a_1 = 3.5 \quad \Leftrightarrow \quad a_1 = 1 \tag{6.26}$$

$$(6.25) \quad \Leftrightarrow \quad 2.5\,a_2 = -2.5 \quad \Leftrightarrow \quad a_2 = -1 \tag{6.27}$$

So, multiple linear regression yields the function

$$x_k^{(1)} \approx 2 + (x_k^{(2)} - 2) - (x_k^{(3)} - 0) = x_k^{(2)} - x_k^{(3)} \tag{6.28}$$

Inserting X (6.18) into (6.28) yields an approximation error of zero in this case. This is usually not the case: the approximation error is usually larger than zero.

Using the pseudo inverse approach we obtain

$$Y = \begin{pmatrix} 6-2 \\ 2-2 \\ 0-2 \\ 0-2 \\ 2-2 \end{pmatrix} = \begin{pmatrix} 4 \\ 0 \\ -2 \\ -2 \\ 0 \end{pmatrix} \quad X = \begin{pmatrix} 4-2 & -2-0 \\ 1-2 & -1-0 \\ 0-2 & 0-0 \\ 1-2 & 1-0 \\ 4-2 & 2-0 \end{pmatrix} = \begin{pmatrix} 2 & -2 \\ -1 & -1 \\ -2 & 0 \\ -1 & 1 \\ 2 & 2 \end{pmatrix} \tag{6.29}$$

and so

$$A = (X^T \cdot X)^{-1} \cdot X^T \cdot Y$$

$$= \left(\begin{pmatrix} 2 & -1 & -2 & -1 & 2 \\ -2 & -1 & 0 & 1 & 2 \end{pmatrix} \begin{pmatrix} 2 & -2 \\ -1 & -1 \\ -2 & 0 \\ -1 & 1 \\ 2 & 2 \end{pmatrix} \right)^{-1} \begin{pmatrix} 2 & -1 & -2 & -1 & 2 \\ -2 & -1 & 0 & 1 & 2 \end{pmatrix} \begin{pmatrix} 4 \\ 0 \\ -2 \\ -2 \\ 0 \end{pmatrix}$$

$$= \begin{pmatrix} 14 & 0 \\ 0 & 10 \end{pmatrix}^{-1} \begin{pmatrix} 14 \\ -10 \end{pmatrix} = \begin{pmatrix} \frac{1}{14} & 0 \\ 0 & \frac{1}{10} \end{pmatrix} \begin{pmatrix} 14 \\ -10 \end{pmatrix} = \begin{pmatrix} 1 \\ -1 \end{pmatrix} \tag{6.30}$$

which corresponds to the solution $a_1 = 1$, $a_2 = -1$ reported above.

6.2   Linear Regression with Nonlinear Substitution

Substituting features by nonlinear functions of features enables linear regression to build nonlinear regression models for specific nonlinear functions. For example, the aerodynamic drag on any object moving through the atmosphere increases quadratically with the velocity. Although the dependence is nonlinear, we can estimate the drag coefficients by linear regression using drag and velocity data. To do so, we compute the *quadratic* velocities from the original velocity data and use the computed quadratic velocity as input feature. Another example for linear regression with nonlinear substitutions is the *polynomial regression* of degree $q \in \{1, 2, \ldots\}$. In polynomial regression we use data for the input x and the output y, and compute the additional input features

$$x^2, \ldots, x^q \tag{6.31}$$

Linear regression with the input x and the $q - 1$ additional computed features then yields the polynomial coefficients $a_0, a_1, \ldots, a_p$ for the function

$$y \approx f(x) = \sum_{i=0}^{p} a_i x^i \tag{6.32}$$

6.3 Robust Regression

The average quadratic error functional in conventional linear regression is very sensitive to inliers and outliers because these have a quadratic impact on the error. *Robust* error functionals aim to reduce the influence of inliers and outliers. Linear regression with robust error functionals is called *robust linear regression*. One example of a robust error functional is the Huber function where the errors are only squared if their absolute values are smaller than a threshold $\varepsilon > 0$, otherwise they have only a linear effect, and the linear parts are chosen so that the Huber function is continuously differentiable at $\pm\varepsilon$ [6].

$$E_H = \sum_{k=1}^{n} \begin{cases} e_k^2 & \text{if } |e_k| < \varepsilon \\ 2\varepsilon \cdot |e_k| - \varepsilon^2 & \text{otherwise} \end{cases} \tag{6.33}$$

Another example of a robust error functional is *least trimmed squares* [11] which sorts the errors so that

$$e_1' \le e_2' \le \ldots \le e_n' \tag{6.34}$$

and only considers the data with the m smallest errors, $1 \le m \le n$.

$$E_{LTS} = \sum_{k=1}^{m} e_k'^2 \tag{6.35}$$

Changing the considered data may change the regression model and vice versa, so regression and trimming may be done alternatingly several times, for example until the iteration reaches a fixed point when the considered data do not change any more.

6.4 Neural Networks

A popular class of nonlinear regression methods are the so-called *universal approximators* [5, 7]. Consider any continuous real-valued function f on any compact subset $U \subset \mathbb{R}^n$

$$f : U \to \mathbb{R} \tag{6.36}$$

A class F of such functions f is called a *universal approximator* if and only if for any $\epsilon > 0$ there exists a function $f^* \in F$ such that

$$|f(x) - f^*(x)| < \epsilon \tag{6.37}$$

for all $x \in U$. Linear regression or polynomial regression with fixed degree does not yield universal approximators, because data samples from higher order functions can be constructed for which no lower order function can be constructed with an approximation error below a certain nonzero limit, e.g. when trying to approximate points from a square function using a linear function. However, polynomial regression with arbitrary degree does yield universal approximators, because increasing the polynomial degree may reduce the error until it falls below a given threshold. Further examples for universal approximators can be constructed by allowing variable complexity of the regression functions, like sums of arbitrary many functions, or arbitrarily nested functions. This section and the next section present two other examples for such universal approximators: the *multilayer perceptron* and *radial basis function networks*.

A *multilayer perceptron (MLP)* [10] is a *feedforward neural network* [4] that can be represented as a directed graph as shown in Fig. 6.2 (left). Because of the similarity with biological neural networks, MLP nodes are called *neurons*. The neurons are arranged in a layer structure. The inputs to the neurons in the first layer (input layer) are the network inputs and the outputs of the neurons in the last layer (output layer) are the network outputs. Layers between the input and output layers are called *hidden* layers. The network in Fig. 6.2 (left) has three layers: one input layer, one hidden layer, and one output layer. Each network edge represents a scalar information flow, so the number p of input neurons and the number q of output neurons specifies the input and output dimensionality of the network which then realizes a function with p inputs and q outputs, $f : \mathbb{R}^p \to \mathbb{R}^q$. The example in Fig. 6.2 (left) realizes a function with three inputs and three outputs, $f : \mathbb{R}^3 \to \mathbb{R}^3$. Neurons of adjacent layers may be connected by directed edges. Each directed edge from neuron i to neuron j has a real-valued weight $w_{ij} \in \mathbb{R}$. For simplicity, non-existing edges may be represented by edges with zero weights. The output of each neuron i is the real-valued scalar $O_i \in \mathbb{R}$. The "effective" input to each neuron is the

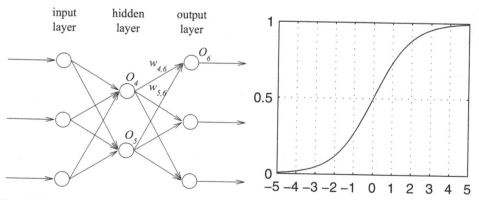

Fig. 6.2 Multilayer perceptron and sigmoidal function

weighted sum of the inputs plus a bias $b_i \in \mathbb{R}$.

$$I_i = \sum_j w_{ji} O_j + b_i \tag{6.38}$$

The output of each neuron is computed from the effective input using a nonlinear *activation function* $s : \mathbb{R} \to \mathbb{R}$, so

$$O_i = s(I_i) \tag{6.39}$$

Frequently used activation functions include the *sigmoid* (s-shaped) functions, for example the *logistic* function

$$s(x) = \frac{1}{1 + e^{-x}} \in (0, 1) \tag{6.40}$$

or the *hyperbolic tangent* function

$$s(x) = \tanh x \in (-1, 1) \tag{6.41}$$

The purpose of the sigmoid functions is to introduce nonlinearity but also to restrict the neuron outputs to a limited range (see the data transformations in Chap. 3). For a given input, the MLP output can be computed using Eqs. (6.38), (6.39), and (6.40) or (6.41). The MLP realizes a function $f : \mathbb{R}^p \to \mathbb{R}^q$ specified by the weights w_{ij} and the bias values b_i. In nonlinear regression, a set $Z = \{(x_1, y_1), \ldots, (x_n, y_n)\} \subset \mathbb{R}^{p+q}$ of pairs of input and output data vectors is used to estimate the weights w_{ij} and the bias values b_i, so that for each input vector x_k, $k = 1, \ldots, n$, the MLP estimates the corresponding output vector y_k, $y_k \approx f(x_k)$, $(x_k, y_k) \in Z$. In the context of artificial neural networks, the estimation of the parameters w_{ij} and b_i from Z is called *training*, and Z is called the *training data set*. To illustrate the MLP training process we first consider three layer MLPs: MLPs with one input layer (p neurons), one hidden layer (h neurons), and one output layer (q neurons), so the MLP has a total of $p + h + q$ neurons and bias values, and a total of $p \cdot h + h \cdot q$ edge weights, where missing connections are represented by zero weights. We define the neuron indices so that neurons $1, \ldots, p$ are the input neurons, $p + 1, \ldots, p + h$ are the hidden neurons, and $p + h + 1, \ldots, p + h + q$ are the output neurons, so the network input is $x = (I_1, \ldots, I_p) \in \mathbb{R}^p$ and the network output is $f(x) = (O_{p+h+1}, \ldots, O_{p+h+q}) \in \mathbb{R}^q$. For a given input x_k the network output will be $f(x_k)$ but it should be y_k, so we want to minimize the average quadratic error between $f(x_k)$ and y_k. We denote $y_k = (O'_{p+h+1}, \ldots, O'_{p+h+q})$ and write the error as

$$E = \frac{1}{r} \cdot \sum_{i=p+h+1}^{p+h+q} (O_i - O'_i)^2 \tag{6.42}$$

We use gradient descent to find the weights w_{ij}, so after initialization we iteratively update the new weights w_{ij} as

$$w_{ij} - \alpha(t) \cdot \frac{\partial E}{\partial w_{ij}} \qquad (6.43)$$

In the neural network context the step length α is usually called *learning rate*. Using the chain rule, the error gradients can be computed from (6.38), (6.39), and (6.42).

$$\frac{\partial E}{\partial w_{ij}} = \frac{\partial E}{\partial O_j} \cdot \frac{\partial O_j}{\partial I_j} \cdot \frac{\partial I_j}{\partial w_{ij}} \qquad (6.44)$$

$$\sim \underbrace{(O_j - O'_j) \cdot s'(I_j)}_{= \delta_j^{(O)}} \cdot \ O_i$$

The term $\delta_j^{(O)}$, $j = p + h + 1, \ldots, p + h + q$, is called the *delta value* of the output layer. With this delta value, the update rule (6.43) can be written as the so-called *delta rule* [12]

$$w_{ij} - \alpha(t) \cdot \delta_j^{(O)} \cdot O_i \qquad (6.45)$$

For the logistic function (6.40), for example, we can write the derivative $s'(I_j)$ as

$$s'(I_j) = \frac{\partial}{\partial I_j} \frac{1}{1 + e^{-I_j}} = -\frac{1}{(1 + e^{-I_j})^2} \cdot \left(-e^{-I_j}\right)$$

$$= \frac{1}{1 + e^{-I_j}} \cdot \frac{e^{-I_j}}{1 + e^{-I_j}} = O_j \cdot (1 - O_j) \qquad (6.46)$$

so (6.44) and (6.46) yield a mathematically elegant formula for the delta value for the logistic activation function.

$$\delta_j^{(O)} = (O_j - O'_j) \cdot O_j \cdot (1 - O_j) \qquad (6.47)$$

To update the weights between input layer and hidden layer we have to consider the sum of derivatives along all paths.

$$
\begin{aligned}
\frac{\partial E}{\partial w_{ij}} &= \sum_{l=p+h+1}^{p+h+q} \frac{\partial E}{\partial O_l} \cdot \frac{\partial O_l}{\partial I_l} \cdot \frac{\partial I_l}{\partial O_j} \cdot \frac{\partial O_j}{\partial I_j} \cdot \frac{\partial I_j}{\partial w_{ij}} \\
&\sim \sum_{l=p+h+1}^{p+h+q} (O_l - O_l') \cdot s'(I_l) \cdot w_{jl} \cdot s'(I_j) \cdot O_i \qquad (6.48) \\
&= \sum_{l=p+h+1}^{p+h+q} \delta_l^{(O)} \qquad w_{jl} \cdot s'(I_j) \cdot O_i
\end{aligned}
$$

In correspondence with (6.44) we find that

$$
\frac{\partial E}{\partial w_{ij}} = s'(I_j) \cdot \underbrace{\sum_{l=p+h+1}^{p+h+q} \delta_l^{(O)} \cdot w_{jl} \cdot O_i}_{= \delta_j^{(H)}} \qquad (6.49)
$$

so corresponding to (6.45), the delta rule for the new weights between the input layer and the hidden layer is

$$
w_{ij} - \alpha(t) \cdot \delta_j^{(H)} \cdot O_i \qquad (6.50)
$$

For the logistic activation function we obtain with (6.46) and (6.49)

$$
\delta_j^{(H)} = (O_j) \cdot (1 - O_j) \cdot \sum_{l=p+h+1}^{p+h+q} \delta_l^{(O)} \cdot w_{jl} \qquad (6.51)
$$

To derive the update rules for the bias values b_i we proceed in the same way, except that we have $\partial I_j / \partial b_i = 1$, so the last factor O_i disappears in (6.44), (6.45), (6.48)–(6.50). Notice the correspondence of the delta rules for the different layers. For more than three layers, we obtain corresponding delta rules, instances of the so-called *generalized delta rule*, where the delta values are computed from the neuron outputs of the corresponding layer and the delta values of the succeeding layers. The weights of the whole network can be efficiently updated by starting with the weights of the output layer and then propagating back layer by layer up to the input layer. This scheme is called the *backpropagation algorithm* [15]. The backpropagation algorithm iteratively updates the MLP weights for each input-output vector of the training data set, and passes through the training data several times, where each pass is called a learning *epoch*. The complete backpropagation algorithm (for a three layer MLP) is illustrated in Fig. 6.3. After training the MLP can be

1. input: neuron numbers $p, h, q \in \{1, 2, \ldots\}$,
   input training data $X = \{x_1, \ldots, x_n\} \subset \mathbb{R}^p$,
   output training data $Y = \{y_1, \ldots, y_n\} \subset \mathbb{R}^q$,
   learning rate $\alpha(t)$
2. initialize weights w_{ij} and biases b_j for $i = 1, \ldots, p$, $j = p+1, \ldots, p+h$
   and for $i = p+1, \ldots, p+h$, $j = p+h+1, \ldots, p+h+q$
3. for each input-output vector pair (x_k, y_k), $k = 1, \ldots, n$

   a. update the weights and biases of the output layer
   $$w_{ij} = w_{ij} - \alpha(t) \cdot \delta_j^{(O)} \cdot O_i, \quad \begin{array}{l} i = p+1, \ldots, p+h \\ j = p+h+1, \ldots, p+h+q \end{array}$$
   $$b_j = b_j - \alpha(t) \cdot \delta_j^{(O)}, \quad j = p+h+1, \ldots, p+h+q$$
   b. update the weights and biases of the hidden layer
   $$w_{ij} = w_{ij} - \alpha(t) \cdot \delta_j^{(H)} \cdot O_i, \quad \begin{array}{l} i = 1, \ldots, p \\ j = p+1, \ldots, p+h \end{array}$$
   $$b_j = b_j - \alpha(t) \cdot \delta_j^{(H)}, \quad j = p+1, \ldots, p+h$$

4. repeat from (3.) until termination criterion holds
5. output: weights w_{ij} and biases b_j for $i = 1, \ldots, p$, $j = p+1, \ldots, p+h$
   and for $i = p+1, \ldots, p+h$, $j = p+h+1, \ldots, p+h+q$

Fig. 6.3 Backpropagation algorithm for three layer MLP

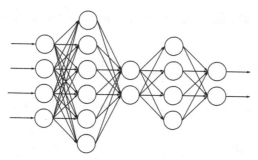

Fig. 6.4 A deep neural network structure

used as a nonlinear regression function for previously unknown input vectors, which is called the *recall* mode. Recently, MLPs with much more than three layers have become popular, so-called *deep neural networks* [2], as illustrated in Fig. 6.4. In deep networks usually only a subset of the network layers is trained at a time and then frozen, so the layer structure is iteratively trained, often using the auto-encoder architecture presented in Sect. 6.4.

6.5 Radial Basis Function Networks

As a second example of universal approximators we present *radial basis function (RBF) networks* [9]. Each RBF neuron implements a single radial basis function with the argument $\|x - \mu_i\|$, for example the frequently used Gaussian function

$$u_i(x) = e^{-\left(\frac{\|x - \mu_i\|}{\sigma_i}\right)^2} \tag{6.52}$$

$\mu_i \in \mathbb{R}^p, \sigma_i > 0, i = 1, \ldots, c$. Figure 6.5 shows an RBF network with three neurons and, correspondingly, three basis functions. Each neuron is connected with the input $x \in \mathbb{R}^p$ and, via a weight $w_i \in \mathbb{R}$, with the additive output neuron. Notice that each edge can carry a vector here as opposed to the scalar edges for MLPs in the previous section. The output of a Gaussian RBF network is

$$y = \sum_{i=1}^{c} w_i \cdot e^{-\left(\frac{\|x - \mu_i\|}{\sigma_i}\right)^2} \tag{6.53}$$

An RBF network realizes a nonlinear function as a weighted sum of radial basis functions. RBF networks have two sets of parameters: the RBF parameters (the centers $\mu_1, \ldots, \mu_c$ and standard deviations $\sigma_1, \ldots, \sigma_c$ for Gaussian RBFs) and the weights $w_1, \ldots, w_c$. The RBF parameters may be found by clustering (see Chap. 9). Here, we illustrate an *alternating optimization* scheme, where the RBF parameters and weights are alternatingly optimized. The output of the RBF network is a linear combination of the c individual RBF functions, so the weight coefficients can be found by linear regression, for example using the pseudo inverse approach. The RBF parameters may be found by gradient descent. The

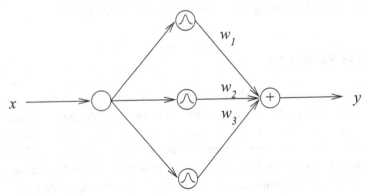

Fig. 6.5 RBF network with three radial basis functions

average quadratic error of the RBF network is

$$E = \frac{1}{n} \sum_{k=1}^{n} \left(\sum_{i=1}^{c} w_i \, e^{-\left(\frac{\|x_k - \mu_i\|}{\sigma_i}\right)^2} - y_k \right)^2 \tag{6.54}$$

This yields the error gradients

$$\frac{\partial E}{\partial \mu_i} = \frac{4 w_i}{n \sigma_i^2} \sum_{k=1}^{n} \left(\sum_{j=1}^{m} w_j \, e^{-\left(\frac{\|x_k - \mu_j\|}{\sigma_j}\right)^2} - y_k \right) \|x_k - \mu_i\| \, e^{-\left(\frac{\|x_k - \mu_i\|}{\sigma_i}\right)^2} \tag{6.55}$$

$$\frac{\partial E}{\partial \sigma_i} = \frac{4 w_i}{n \sigma_i^3} \sum_{k=1}^{n} \left(\sum_{j=1}^{m} w_j \, e^{-\left(\frac{\|x_k - \mu_j\|}{\sigma_j}\right)^2} - y_k \right) \|x_k - \mu_i\|^2 \, e^{-\left(\frac{\|x_k - \mu_i\|}{\sigma_i}\right)^2} \tag{6.56}$$

To find the vector $W = (w_1 \dots w_c)^T$ of weights for the given target outputs $Y = (y_1 \dots y_n)^T$ we write the outputs of the hidden layer as a matrix

$$U = \begin{pmatrix} e^{-\left(\frac{x_1 - \mu_1}{\sigma_1}\right)^2} & \cdots & e^{-\left(\frac{x_1 - \mu_c}{\sigma_c}\right)^2} \\ \vdots & & \vdots \\ e^{-\left(\frac{x_n - \mu_1}{\sigma_1}\right)^2} & \cdots & e^{-\left(\frac{x_n - \mu_c}{\sigma_c}\right)^2} \end{pmatrix} \tag{6.57}$$

and obtain by linear regression using the pseudo inverse

$$Y = U \cdot W \quad \Rightarrow \quad W = (U^T \cdot U)^{-1} \cdot U^T \cdot Y \tag{6.58}$$

The corresponding RBF training algorithm is summarized in Fig. 6.6.

6.6   Cross-Validation

Cross-validation is a method to validate models generated from data, for example regression models. Nonlinear regression models are often able to model given training data with very low or even zero error, if sufficiently many model parameters are used, for example when the number of hidden layer neurons or RBF functions is very high. Very low error models tend to *overfitting*, which means that they accurately match the training data rather than the underlying input-output relation and therefore do not generalize well to unknown data. Such overfitting can be avoided by cross-validation. The idea of cross-validation is to partition the available input-output data set $Z = \{(x_1, y_1), \dots, (x_n, y_n)\} \subset \mathbb{R}^{p+q}$ into a training data set $Z_t \subset Z$ and a disjoint validation data set $Z_v \subset Z$, $Z_t \cap Z_v = \{\}$, $Z_t \cup Z_v = Z$, where the training data are used to build (train) a model f, and the validation

1. input: RBF number $c \in \{1, 2, \ldots\}$,
   input training data $X = \{x_1, \ldots, x_n\} \subset \mathbb{R}^p$,
   output training data $Y = \{y_1, \ldots, y_n\} \subset \mathbb{R}^r$,
   learning rate $\alpha(t)$
2. initialize weights $w_1, \ldots, w_c \in \mathbb{R}$
3. for each input-output vector pair (x_k, y_k), $k = 1, \ldots, n$,
   update the RBF parameters

$$\mu_i = \mu_i - \alpha(t) \cdot \frac{\partial E}{\partial \mu_i}, \quad i = 1, \ldots, c$$

$$\sigma_i = \sigma_i - \alpha(t) \cdot \frac{\partial E}{\partial \sigma_i}, \quad i = 1, \ldots, c$$

   until termination criterion holds
4. compute matrix of hidden layer outputs

$$U = \begin{pmatrix} e^{-\left(\frac{x_1 - \mu_1}{\sigma_1}\right)^2} & \cdots & e^{-\left(\frac{x_1 - \mu_c}{\sigma_c}\right)^2} \\ \vdots & & \vdots \\ e^{-\left(\frac{x_n - \mu_1}{\sigma_1}\right)^2} & \cdots & e^{-\left(\frac{x_n - \mu_c}{\sigma_c}\right)^2} \end{pmatrix}$$

5. compute optimal weights $w_1, \ldots, w_c \in \mathbb{R}$

$$W = (U^T \cdot U)^{-1} \cdot U^T \cdot Y$$

6. repeat from (3.) until termination criterion holds
7. output: RBF parameters μ_i, σ_i and weights w_i, $i = 1, \ldots, c$

Fig. 6.6 Training algorithm for RBF networks

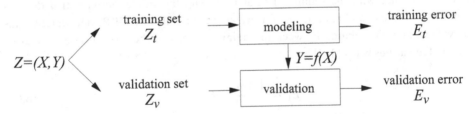

Fig. 6.7 Cross-validation

data are used to validate this trained model (Fig. 6.7). For regression problems, the average quadratic training error is

$$E_t = \frac{1}{|Z_t|} \sum_{(x,y) \in Z_t} \|y - f(x)\|^2 \tag{6.59}$$

and the average quadratic validation error is

$$E_v = \frac{1}{|Z_v|} \sum_{(x,y)\in Z_v} \|y - f(x)\|^2 \qquad (6.60)$$

Since the validation data are not used for training, a low validation error indicates that the trained model is a good representation of the underlying input-output relation.

k-fold cross-validation randomly partitions Z into k pairwise disjoint and (almost) equally sized subsets $Z_1, \ldots, Z_k$, where $Z_i \cap Z_j = \{\}$ for all $i \neq j$, $Z_1 \cup \ldots \cup Z_k = Z$, and $|Z_i| \approx |Z_j|$ for all $i, j = 1, \ldots, k$. Each of the subsets Z_i, $i = 1, \ldots, k$, is used to validate the model f_i trained with the remaining $k-1$ subsets Z_j, $j = 1, \ldots, k$, $j \neq i$. This yields k validation errors $E_{v1}, \ldots, E_{vk}$ for the k models $f_1, \ldots, f_k$

$$E_{vi} = \frac{1}{|Z_i|} \sum_{(x,y)\in Z_i} \|y - f_i(x)\|^2 \qquad (6.61)$$

$i = 1, \ldots, k$, whose average is called the *k-fold cross-validation error*.

$$E_v = \frac{1}{k} \sum_{i=1}^{k} E_{vi} \qquad (6.62)$$

Notice that this is the average error of k *different* models, so cross-validation does not assess the quality of one individual model but of the modeling method and its parameters, for example a three-layer MLP with five hidden neurons.

Leave one out is a cross-validation scheme where for each model only one single data vector is retained for validation. So, leave one out can be viewed as n-fold cross-validation.

The validation error is typically larger than the training error, $E_v > E_t$, because the models are trained with the training data and not with the validation data. If d denotes the number of free model parameters, for example the number of MLP edge weights plus bias values, then the relation between the training error and the validation error can be estimated in various ways [1, 3] as

$$E_v \approx \frac{1 + d/n}{1 - d/n} E_t \qquad (6.63)$$

$$E_v \approx (1 + 2d/n) E_t \qquad (6.64)$$

$$E_v \approx \frac{1}{(1 - d/n)^2} E_t \qquad (6.65)$$

Models are good generalizations when the validation error is close to the training error, $E_v \approx E_t$. For the three estimates above this implies $d \ll n$. Suitable values of d may be obtained by either starting with low d and successively increasing d until training and validation errors start to diverge too much, or by starting with high d and successively decreasing d until training and validation errors become similar.

A well trained and validated regression model may be interpreted as a good approximation of the underlying function between inputs and outputs. If the validation error is low, then inputs and outputs correlate well. If the validation error is high, then inputs and outputs do not correlate well. Thus, the inverse of the validation error can be viewed as a correlation measure between the model inputs and outputs, as an alternative to the correlation measures presented in the previous chapter.

6.7 Feature Selection

In the previous sections we assumed that the output features in Y depend on all input features in X, and therefore all features in X were considered for the (regression) model. However, in real-world applications this is often not the case. Some of the features in X may be irrelevant for Y. Taking into account irrelevant features may lead to unnecessarily high model complexity (too many model parameters and high computational effort for training) or even to lower model performance, for example if the model overfits because of irrelevant features. Therefore, in many data analysis tasks it is useful to select and use only the relevant features, not only in regression but also in forecasting, classification and clustering (see the following three chapters). In Chap. 8 we will present a decision tree approach that implicitly finds the most relevant features for a given classification task. However, any regression, forecasting, classification and clustering method can be combined with feature selection. To do so, subsets of the available features are used for model building and are successively adapted according to the model performance until a satisfactory feature subset is found. A simple feature selection approach starts with the complete set of features and successively removes the least relevant feature until the model performance falls below a given threshold. The reverse approach starts with an empty feature set and successively adds the most relevant feature until no significant model improvement can be achieved. Because of possibly complex interdependencies between the features both approaches may find suboptimal feature subsets. For example, if two features *individually* have a low relevance but jointly have a high relevance, then the second approach will not take into account either of them. This drawback can be reduced by combining both approaches where the feature sets are alternatingly extended and reduced. To find a set of features which are optimal in some well-defined sense, all possible combinations need to be evaluated. This is not feasible since the number of possible feature subsets increases exponentially with the number of features. Many stochastic methods have been proposed to *approximate* an optimal feature selection, including meta-heuristics such as *simulated annealing* [8], *evolutionary algorithms* [13], or *swarm intelligence* [14]. Notice that feature selection yields (axis-parallel) projections of the data (see Chap. 4).

Problems

6.1 Consider the data sets $X = \{1, 2, 4, 5\}$, $Y = \{-1, 1, 1, -1\}$.

(a) Which function $y = f(x)$ will be found by linear regression?
(b) When we add an outlier at $x_5 = 3$, $y_5 = 15$, which function $y = f(x)$ will then be found by linear regression?
(c) Now we use robust linear regression, $\varepsilon = 4$, with the error functional

$$E_H = \frac{1}{n} \sum_{k=1}^{n} \begin{cases} e_k^2 & \text{if } |e_k| < \varepsilon \\ 2\varepsilon \cdot |e_k| - \varepsilon^2 & \text{otherwise} \end{cases}$$

Which outlier value y_5' would have the same effect as the outlier value y_5 in (b)?
(d) How do you interpret these results?

6.2 Sketch an MLP with hyperbolic tangent transfer functions and possible additional constant (bias) inputs at each neuron so that the network (approximately) realizes

(a) a hyperbolic tangent function
(b) a cosine function for inputs $x \in [-\pi, \pi]$
(c) an XOR function with inputs x_1, x_2, and output y, so that

x_1	-1	-1	$+1$	$+1$
x_2	-1	$+1$	-1	$+1$
y	-1	$+1$	$+1$	-1

6.3 Consider an MLP with *linear* transfer functions (slope one, offset zero), input layer (neurons 1, 2, and 3), hidden layer (neurons 4 and 5), and output layer (neurons 6, 7, and 8), and the weight matrix

$$W = \begin{pmatrix} 0 & 0 & 0 & 1 & 1 & 0 & 0 & 0 \\ 0 & 0 & 0 & 1 & -1 & 0 & 0 & 0 \\ 0 & 0 & 0 & 0 & 0 & 0 & 0 & 0 \\ 0 & 0 & 0 & 0 & 0 & a & b & c \\ 0 & 0 & 0 & 0 & 0 & d & e & f \end{pmatrix}$$

with real valued parameters a, b, c, d, e, f.

(a) Which function does this MLP realize?
(b) How would you choose the parameters a, b, c, d, e, f so that this MLP can be used as an auto-encoder?
(c) What are the advantages and disadvantages of this auto-encoder?

6.4 What does a small training error and a large validation error indicate?

References

1. H. Akaike. A new look at the statistical model identification. *IEEE Transactions on Automatic Control*, AC–19:716–723, 1974.
2. Y. Bengio. Learning deep architectures for AI. *Foundations and Trends in Machine Learning*, 2(1):1–127, 2009.
3. P. Craven and G. Wahba. Smoothing noisy data with spline functions: Estimating the correct degree of smoothing by the method of generalized cross–validation. *Numerical Mathematics*, 31:377–403, 1979.
4. R. Hecht-Nielsen. *Neurocomputing*. Addison-Wesley, 1990.
5. K. Hornik, M. Stinchcombe, and H. White. Multilayer feedforward networks are universal approximators. *Neural Networks*, 2(5):359–366, 1989.
6. P. J. Huber. *Robust Statistics*. Wiley, New York, 2nd edition, 2009.
7. A. N. Kolmogorov. On the representation of continuous functions of many variables by superposition of continuous functions of one variable and addition. *Doklady Akademii Nauk SSSR*, 144:679–681, 1957.
8. R. Meiria and J. Zahavi. Using simulated annealing to optimize the feature selection problem in marketing applications. *European Journal of Operational Research*, 171(3):842–858, 2006.
9. J. Moody and C. Darken. Fast learning in networks of locally-tuned processing units. *Neural Computation*, 1:281–294, 1989.
10. F. Rosenblatt. The perceptron: A probabilistic model for information storage and organization in the brain. *Psychological Reviews*, 65:386–408, 1958.
11. P. J. Rousseeuw and A. M. Leroy. *Robust Regression and Outlier Detection*. Wiley, New York, 1987.
12. D. E. Rumelhart, G. E. Hinton, and R. J. Williams. Learning internal representations by error backpropagation. In D. E. Rumelhart and J. L. McClelland, editors, *Parallel Distributed Processing. Explorations in the Microstructure of Cognition*, volume 1, pages 318–362. MIT Press, Cambridge, 1986.
13. W. Siedlecki and J. Sklansky. A note on genetic algorithms for large–scale feature selection. *Pattern Recognition Letters*, 10(5):335–347, 1989.
14. S. Vieira, J. M. Sousa, and T. A. Runkler. Multi-criteria ant feature selection using fuzzy classifiers. In C. A. Coello Coello, S. Dehuri, and S. Ghosh, editors, *Swarm Intelligence for Multi-objective Problems in Data Mining*, pages 19–36. Springer, 2009.
15. P. J. Werbos. *The Roots of Backpropagation: From Ordered Derivatives to Neural Networks and Political Forecasting (Adaptive and Learning Systems for Signal Processing, Communications and Control Series)*. Wiley–Interscience, 1994.

Forecasting

7

Abstract

For forecasting future values of a time series we imagine that the time series is generated by a (possibly noisy) deterministic process such as a Mealy or a Moore machine. This leads to recurrent or auto-regressive models. Building forecasting models is essentially a regression task. The training data sets for forecasting models are generated by finite unfolding in time. Popular linear forecasting models are auto-regressive models (AR) and generalized AR models with moving average (ARMA), with integral terms (ARIMA), or with local regression (ARMAX). Popular nonlinear forecasting models are recurrent neural networks.

7.1 Finite State Machines

Sequences of data play an important role in data analysis. In Chap. 2 we presented relations for sequential data and sampling schemes for continuous signals. In Chap. 3 we showed how sequences can be preprocessed by filtering. And in Chap. 4 we illustrated how sequential data can be analyzed and visualized using spectral analysis. In data analysis we not only preprocess, visualize and analyze sequential data but also try to use observed data sequences to produce reliable forecasts of future data.

Finding an appropriate forecasting model begins with the assumption that the sequence is generated by a dynamic feedback system [2] with an input $x \in \mathbb{R}^p$, a hidden state $s \in \mathbb{R}^h$, and an output $y \in \mathbb{R}^q$. Notice the similarity to the three-layer MLP architecture presented in the previous chapter. At each time step k such a dynamic feedback system can be described by a state equation

$$s_k = f_s(s_{k-1}, x_k) \tag{7.1}$$

© Springer Fachmedien Wiesbaden GmbH, part of Springer Nature 2020
T. A. Runkler, *Data Analytics*, https://doi.org/10.1007/978-3-658-29779-4_7

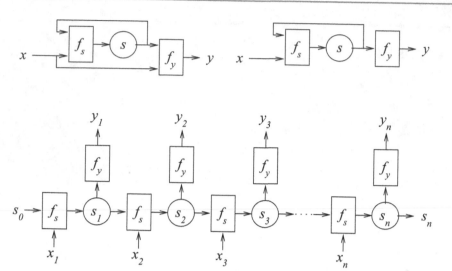

Fig. 7.1 Finite state machines: Mealy machine (top left), Moore machine (top right), unfolded Moore machine (bottom)

and an output equation

$$y_k = f_y(s_k, x_k) \tag{7.2}$$

The top left view of Fig. 7.1 shows a block diagram of such a system. If the set of states that can be reached in a such system is finite, we call the system a *finite state machine*, for example when $s \in \{0, 1\}^r$, $r \in \{1, 2, \ldots\}$. Equations (7.1) and (7.2) describe a so-called *Mealy machine* [3]. Special cases of Mealy machines are *Moore machines* [4], where the output only depends on the state but not the current input. So a Moore machine is described by a state equation

$$s_k = f_s(s_{k-1}, x_k) \tag{7.3}$$

and an output equation

$$y_k = f_y(s_k) \tag{7.4}$$

The top right view of Fig. 7.1 shows a block diagram of a Moore machine. Any Mealy machine can be translated into an equivalent Moore machine and vice versa. To realize a given behavior, Mealy machines usually require a smaller state space. However, Moore machines are often easier to design and their behavior is easier to analyze. Without loss of generality we will restrict here to Moore machines.

If we distinguish the inputs x, states s, and outputs y for the time steps $1, \ldots, n$, then we obtain the so-called unfolded representation of a Moore machine that is shown in the bottom view of Fig. 7.1.

7.2 Recurrent Models

Forecasting uses the observed input and output sequences $(x_1, \ldots, x_n)$ and $(y_1, \ldots, y_n)$ to predict the expected output sequence $(y_{n+1}, \ldots, y_{n+q}\}$ without knowing the expected input sequence $(x_{n+1}, \ldots, x_{n+q}\}$. To build a forecasting model we have to find functions f_s and f_y that approximate the observed data in some well-defined sense, for example by minimizing the average quadratic error between the observations y_k and the predicted values y'_k.

$$E = \frac{1}{n} \sum_{k=1}^{n} (y_k - y'_k)^2 \qquad (7.5)$$

This leads to a regression problem (just as described in the previous chapter), where the functions f_s and f_y are estimated from the quadruples $(x_k, s_{k-1}, s_k, y_k), k = 1, \ldots, n$. The internal states s are usually unknown and have to be estimated from the observed data. Each state s_k depends on the (also unknown) previous states $s_0, \ldots, s_{k-1}$. The previous states were influenced by the input values $x_1, \ldots, x_{k-1}$ via the function f_s and have impacted the output values $y_1, \ldots, y_{k-1}$ via the function f_y. We can therefore assume that s_k can be estimated from $x_1, \ldots, x_{k-1}$ and $y_1, \ldots, y_{k-1}$. This assumption yields a state free model that can be interpreted as an approximation of a Moore machine:

$$y_k = f_k(y_1, \ldots, y_{k-1}, x_1, \ldots, x_{k-1}), \quad k = 2, \ldots, n \qquad (7.6)$$

The functions $f_2, \ldots, f_n$ cannot be found by regression because at each time step k only a single data tuple is available. This can be overcome when the state is not estimated from all previous input and output values but only from the last m of these corresponding to a moving time window of size m:

$$y_k = f(y_{k-m}, \ldots, y_{k-1}, x_{k-m}, \ldots, x_{k-1}), \quad k = m + 1, \ldots, n \qquad (7.7)$$

The function f can be estimated from $n - m$ data tuples. For example, for $m = 3$ and $n = 8$ these $8 - 3 = 5$ tuples are

$$y_4 = f(y_1, y_2, y_3, x_1, x_2, x_3) \qquad (7.8)$$

$$y_5 = f(y_2, y_3, y_4, x_2, x_3, x_4) \qquad (7.9)$$

1. input: data sets $X = \{x_1, \ldots, x_n\} \subset \mathbb{R}^p$, $Y = \{y_1, \ldots, y_n\} \subset \mathbb{R}^q$, parameter $m \in \{1, \ldots, n\}$
2. construct regression data set Z according to (7.7)
3. estimate f from Z using a suitable regression method
4. output: f

Fig. 7.2 State free forecasting

$$y_6 = f(y_3, y_4, y_5, x_3, x_4, x_5) \tag{7.10}$$

$$y_7 = f(y_4, y_5, y_6, x_4, x_5, x_6) \tag{7.11}$$

$$y_8 = f(y_5, y_6, y_7, x_5, x_6, x_7) \tag{7.12}$$

The values $x_m, \ldots, x_{n-m}$ and $y_m, \ldots, y_{n-m}$ ($x_3, x_4, x_5, y_3, y_4, y_5$ in this example) occur most often, so these influence the forecasting model most. The number of considered time steps m should not be too small to make sure that the hidden states can be estimated well, and it should not be too large to make sure that the number of data tuples is sufficient for a good regression model. Figure 7.2 illustrates the algorithm to build state free forecasting models. This algorithm reduces forecasting to a regression problem. Any regression method may be used here, for example any of the methods presented in Chap. 6. Suitable values for the parameter m and for the parameters of the chosen regression model can be determined by cross-validation (see Chap. 6). The forecasting model yields the model-based estimates of the (already observed) outputs $y_1, \ldots, y_n$, and it can be used to produce forecasts of the future outputs $y_{n+1}, y_{n+2}, \ldots$ if the future inputs $x_{n+1}, x_{n+2}, \ldots$ are known or can at least be estimated. A simple estimation is the average of the previously observed inputs.

$$x_k = \frac{1}{n} \sum_{j=1}^{n} x_j, \quad k = n+1, n+2, \ldots \tag{7.13}$$

7.3 Autoregressive Models

The problem of unknown future inputs can be avoided by models that do not consider the inputs at all, the so-called *auto-regressive models*.

$$y_k = f(y_{k-m}, \ldots, y_{k-1}), \quad k = m+1, \ldots, n \tag{7.14}$$

For example, for $m = 3$ and $n = 8$ the regression tuples are

$$y_4 = f(y_1, y_2, y_3) \tag{7.15}$$

$$y_5 = f(y_2, y_3, y_4) \tag{7.16}$$

$$y_6 = f(y_3, y_4, y_5) \tag{7.17}$$

$$y_7 = f(y_4, y_5, y_6) \tag{7.18}$$

$$y_8 = f(y_5, y_6, y_7) \tag{7.19}$$

For linear regression such autoregressive models are called *linear autoregressive (AR) models* [1]. Such linear models can be estimated very efficiently but they yield suboptimal results if the underlying relations are highly nonlinear. To obtain better results for nonlinear systems, linear AR models use moving averages *(ARMA models)*, integral terms *(ARIMA models)*, or local regression *(ARMAX models)*. More accurate nonlinear forecasting models can be found with nonlinear regression, for example using neural networks, which are then called *recurrent* neural networks [5].

Problems

7.1 We want to predict the sales figures of a startup online shop from the sales of the previous 3 months: 5000 Euros, 10000 Euros, 15000 Euros.

(a) Compute the μ-σ-standardized time series.
(b) Construct the (standardized) regression data set for a forecasting model with time horizon 1 and find the optimal linear autoregressive forecasting model with offset $(f(0) \neq 0)$ and time horizon 1 for these data.
(c) Using this forecasting model compute the (unstandardized) sales forecasts for the next 2 months.
(d) Which value will this linear forecasting model yield if time goes to infinity?

7.2 We construct a *single* layer perceptron (SLP) using the linear forecasting model from Problem 7.1 followed by a single neuron with hyperbolic tangent activation function. If necessary use the following approximations: $\tanh(3/4) = 5/8$, $\tanh(1) = 3/4$, $\tanh(3/2) = 29/32$, $\tanh(7/4) = 15/16$, $\tanh(2) = 1$.

(a) Initialize the SLP with (the standardized equivalent of) 5000 Euros and compute the (unstandardized) sales forecasts for the following 3 months.
(b) Which value will this SLP forecasting model yield if time goes to infinity?

References

1. G. E. P. Box, G. M. Jenkins, and G. C. Reinsel. *Time Series Analysis: Forecasting and Control.* Prentice Hall, 4th edition, 2008.
2. J. E. Hopcroft, R. Motwani, and J. D. Ullman. *Introduction to Automata Theory, Languages, and Computation.* Addison Wesley, 3rd edition, 2006.
3. G. H. Mealy. A method for synthesizing sequential circuits. *Bell System Technology Journal,* 34:1045–1079, September 1955.
4. E. F. Moore. Gedanken experiments on sequential machines. In W. R. Ashby, C. E. Shannon, and J. McCarthy, editors, *Automata studies*, pages 129–156. Princeton University Press, 1956.
5. H. G. Zimmermann and R. Neuneier. Modeling dynamical systems by recurrent neural networks. In *International Conference on Data Mining*, pages 557–566, Cambridge, 2000.

Classification

8

Abstract

Classification can be done by supervised learning that uses labeled data to assign objects to classes. We distinguish false positive and false negative errors and present numerous indicators to quantify classifier performance. Often pairs of indicators are considered to assess classification performance. We illustrate this with the receiver operating characteristic and the precision recall diagram. Several different classifiers with specific capabilities and limitations are presented in detail: the naive Bayes classifier, linear discriminant analysis, the support vector machine (SVM) using the kernel trick, nearest neighbor classifiers, learning vector quantification, and hierarchical classification using decision trees.

8.1 Classification Criteria

The previous chapters have considered two types of data: feature data

$$X = \{x_1, \ldots, x_n\} \subset \mathbb{R}^p \tag{8.1}$$

and pairs of input and output features

$$Z = \{(x_1, y_1), \ldots, (x_n, y_n)\} \subset \mathbb{R}^{p+q} \tag{8.2}$$

This chapter considers data of objects that are assigned to c *classes*, $c \in \{2, 3, \ldots\}$. For example, a medical examination yields features of a patient such as body temperature or blood pressure that can be used to determine whether the patient is healthy or sick, i.e. whether the patient can be assigned to the class of healthy or sick patients. The class

© Springer Fachmedien Wiesbaden GmbH, part of Springer Nature 2020
T. A. Runkler, *Data Analytics*, https://doi.org/10.1007/978-3-658-29779-4_8

assignment of a feature data set can be specified by a class vector $y \subset \{1, \ldots, c\}$. The tuples of feature data and class assignments form a *labeled* feature data set

$$Z = \{(x_1, y_1), \ldots, (x_n, y_n)\} \subset \mathbb{R}^p \times \{1, \ldots, c\} \tag{8.3}$$

Notice that in classification, we denote $y_k \in \{1, \ldots, c\}$ as the class label, and in regression, we denote $y_k \in \mathbb{R}^q$ as the output. Any regression method suitable for discrete outputs y may be used for classification, for example neural networks with discretized output. In this section we consider methods that are explicitly designed for classification.

We assume a systematic relation between the feature vectors and classes of the objects described by a labeled feature data set, and that the objects are representative for their classes. Based on this assumption we use the data to model classes and design classifiers that are able to classify previously unknown objects based on their feature vectors. For example, the data set may indicate that patients with a high body temperature belong to the class of sick patients, so a new patient with a high body temperature would also be classified as sick. So, based on a labeled feature data set, we look for a *classifier* function $f : \mathbb{R}^p \rightarrow \{1, \ldots, c\}$ that yields a class y for a given feature vector x. Constructing a classifier function from data is called *classifier design*, and the application of this function is called *classification* [9, 21]. Classifier design corresponds to regression, and classification corresponds to the application of a regression model to potentially previously unknown feature vectors, like for example in cross-validation. We partition the available data into training and validation data, train classifiers with the training data and validate with the validation data. The goal is to obtain good classification performance on the validation data.

To assess classifier performance we consider a selected class, for example the class of sick patients. An ideal classifier classifies all healthy patients as healthy and all sick patients as sick. An incorrect classification can either mean that a healthy patient is classified as sick, or that a sick patient is classified as healthy. These two types of classification errors may have significantly different meanings and effects and are therefore distinguished. Each classification result belongs to one of the following four cases [2, 15]:

1. *true positive* (TP): $y = i$, $f(x) = i$
   (a sick patient is classified as sick)
2. *true negative* (TN): $y \neq i$, $f(x) \neq i$
   (a healthy patient is classified as healthy)
3. *false positive* (FP): $y \neq i$, $f(x) = i$
   (a healthy patient is classified as sick)
4. *false negative* (FN): $y = i$, $f(x) \neq i$
   (a sick patient is classified as healthy)

FP is also called *type I error* or *error of the first kind*. FN is also called *type II error* or *error of the second kind*. Based on the TP, TN, FP, and FN counts of a training or validation data set, various classification performance criteria are frequently used:

- total number of classifications n=TP+TN+FP+FN
- *true classifications* T=TP+TN
  (number of correctly classified patients)
- *false classifications* F=FP+FN
  (number of incorrectly classified patients)
- *relevance* R=TP+FN
  (number of sick patients)
- *irrelevance* I=FP+TN
  (number of healthy patients)
- *positivity* P=TP+FP
  (number of patients that are classified as sick)
- *negativity* N=TN+FN
  (number of patients that are classified as healthy)
- *true classification rate* or *accuracy* T/n
  (probability that a patient is correctly classified)
- *false classification rate* F/n
  (probability that a patient is incorrectly classified)
- *true positive rate* or *sensitivity* or *recall* or *hit rate* TPR=TP/R
  (probability that a sick patient is classified as sick)
- *true negative rate* or *specificity* TNR=TN/I
  (probability that a healthy patient is classified as healthy)
- *false positive rate* or *fall out* or *false alarm rate* FPR=FP/I
  (probability that a healthy patient is classified as sick)
- *false negative rate* FNR=FN/R
  (probability that a sick patient is classified as healthy)
- *positive predictive value* or *precision* TP/P
  (probability that a sick classified patient is sick)
- *negative predictive value* TN/N
  (probability that a healthy classified patient is healthy)
- *negative false classification rate* FN/N
  (probability that a healthy classified patient is sick)
- *positive false classification rate* FP/P
  (probability that a sick classified patient is healthy)
- *F measure* F=2·TP/(R+P)
  (harmonic mean of precision and recall)

None of these criteria alone is sufficient to assess the performance of a classifier. Good values of a single criterion can often be easily obtained by trivial classifiers. For example,

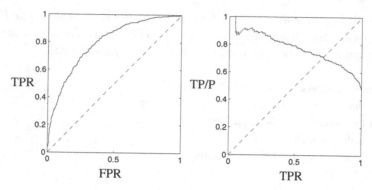

Fig. 8.1 Receiver operating characteristic and precision recall diagram

a true positive rate of 100% (very good) might be achieved by a trivial classifier that always yields the positive class for any arbitrary feature vector. However, such a trivial classifier will also achieve a false positive rate of 100% (very bad). So classifier performance is usually assessed by considering *pairs* of classification criteria.

One example for such an approach is the *receiver operating characteristic (ROC)*, a scatter plot of the true positive rate (TPR) and the false positive rate (FPR). Figure 8.1 (left) shows an example of an ROC diagram. The (training or recall) performance of a certain classifier on a certain data set corresponds to a point in the ROC diagram. Variation of parameters yields an ROC *curve*. The ROC diagram can be used for different purposes, for example to compare training and recall performance, the performance of different classifiers, the effect of different parameter settings, or the performance on different data sets. An optimal classifier yields 100% TPR and 0% FPR which corresponds to the top left corner of the ROC diagram. Therefore a good classification should yield a point close to the top left corner of the ROC diagram. The trivial always positive classifier discussed above yields 100% TPR and 100% FPR (top right corner), and the trivial always negative classifier yields 0% TPR and 0% FPR (bottom left corner). A classifier that always produces the false result yields 0% TPR and 100% FPR (bottom right corner). This could be converted to an optimal classifier by just inverting the classification result. For any classification, inverting the classification result will reflect the corresponding point in the ROC diagram across the center at 50% TPR and 50% FPR. So any classifier that yields a point below the main diagonal (dashed line) can be improved by inverting the classification result. Hence, ROC diagrams usually only show points above the main diagonal.

Another example for assessing classifier performance by considering a pair of classification criteria is the *precision recall (PR) diagram* which displays the precision (or positive predictive value) TP/P versus the recall (true positive rate, sensitivity, or hit rate) TPR. The right view of Fig. 8.1 shows an example of a PR diagram. A high precision can usually be easily achieved for a low recall, i.e. it is easy to build a classifier that correctly classifies only a few sick patients. A higher recall can usually be obtained only with a lower precision. A good classifier maintains a high precision with a high recall. The intersection

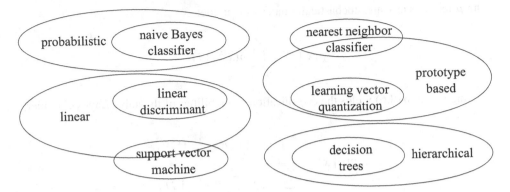

Fig. 8.2 Some important classifier schemes

of the PR curve with the main diagonal is called *precision recall (PR) breakeven point* which is considered an important classification criterion, so a good classifier should have a high PR breakeven point. A detailed discussion and comparison of ROC and PR diagrams can be found in [8].

The following sections present some important families of classifiers and their specific capabilities and limitations: probabilistic, linear, prototype based, and hierarchical classifiers (Fig. 8.2).

8.2 Naive Bayes Classifier

The *naive Bayes classifier* is a probabilistic classifier based on Bayes' theorem [3]: Let A and B denote two probabilistic events, then

$$p(A \mid B) \cdot p(B) = p(B \mid A) \cdot p(A) \tag{8.4}$$

If the event A can be decomposed into the disjoint events $A_1, \ldots, A_c$, $p(A_i) > 0$, $i = 1, \ldots, c$, then

$$p(A_i \mid B) = \frac{p(A_i) \cdot p(B \mid A_i)}{\sum\limits_{j=1}^{c} P(A_j) \cdot P(B \mid A_j)} \tag{8.5}$$

In classification we consider the events "object belongs to class i", or more briefly, just "i", and "object has the feature vector x", or "x". Replacing A_i, A_j and B in (8.5) by these events yields

$$p(i \mid x) = \frac{p(i) \cdot p(x \mid i)}{\sum\limits_{j=1}^{c} p(j) \cdot p(x \mid j)} \tag{8.6}$$

If the p features in x are stochastically independent, then

$$p(x \mid i) = \prod_{k=1}^{p} p(x^{(k)} \mid i) \tag{8.7}$$

Inserting this into (8.6) yields the classification probabilities of the naive Bayes classifier.

$$p(i \mid x) = \frac{p(i) \cdot \prod\limits_{k=1}^{p} p(x^{(k)} \mid i)}{\sum\limits_{j=1}^{c} p(j) \cdot \prod\limits_{k=1}^{p} p(x^{(k)} \mid j)} \tag{8.8}$$

Given a labeled feature data set $Z = \{(x_1, y_1), \ldots, (x_n, y_n)\} \subset \mathbb{R}^p \times \{1, \ldots, c\}$, the prior probabilities can be estimated from the observed frequencies: $p(i)$ is the relative frequency of class $y = i$, $i = 1, \ldots, c$, and the probabilities $p(x^{(k)} \mid i)$ are the relative frequencies of feature $x^{(k)}$ in class $y = i$, $i = 1, \ldots, c$, $k = 1, \ldots, p$. For a given feature vector x, Eq. (8.8) yields the classification probabilities for each class. For a deterministic classification the class with the highest probability is selected.

As an example for naive Bayes classification consider a group of students taking an exam. Some of the students went to class regularly, and the others didn't. Some of the students studied the course material, and the others didn't. Some of the students passed the exam, and the others didn't. The corresponding numbers of students are given in Table 8.1. We want to find the probability that a student will pass the exam, if she or he went to class regularly and studied the course material. From the numbers in Table 8.1 we can compute the probabilities

$$p(\text{went to class} \mid \text{passed}) = \frac{21}{21 + 1} = \frac{21}{22} \tag{8.9}$$

$$p(\text{studied material} \mid \text{passed}) = \frac{16}{16 + 6} = \frac{16}{22} \tag{8.10}$$

$$\Rightarrow p(x \mid \text{passed}) = \frac{21 \cdot 16}{22 \cdot 22} = \frac{84}{121} \tag{8.11}$$

Table 8.1 Naive Bayes classifier: data of the student example

	Number of students who	
	Passed	Failed
Went to class	21	4
Did not go to class	1	3
Studied material	16	2
Did not study material	6	5

$$p(\text{went to class} \mid \text{failed}) = \frac{4}{4+3} = \frac{4}{7} \tag{8.12}$$

$$p(\text{studied material} \mid \text{failed}) = \frac{2}{2+5} = \frac{2}{7} \tag{8.13}$$

$$\Rightarrow p(x \mid \text{failed}) = \frac{4 \cdot 2}{7 \cdot 7} = \frac{8}{49} \tag{8.14}$$

$$p(\text{passed}) = \frac{22}{22+7} = \frac{22}{29} \tag{8.15}$$

$$p(\text{failed}) = \frac{7}{22+7} = \frac{7}{29} \tag{8.16}$$

$$p(\text{passed}) \cdot p(x \mid \text{passed}) = \frac{22}{29} \cdot \frac{84}{121} = \frac{168}{319} \tag{8.17}$$

$$p(\text{failed}) \cdot p(x \mid \text{failed}) = \frac{7}{29} \cdot \frac{8}{49} = \frac{8}{203} \tag{8.18}$$

$$\Rightarrow p(\text{passed} \mid x) = \frac{\frac{168}{319}}{\frac{168}{319} + \frac{8}{203}} \tag{8.19}$$

$$= \frac{168 \cdot 203}{168 \cdot 203 + 8 \cdot 319} = \frac{147}{158} \approx 93\% \tag{8.20}$$

So, a student who went to class regularly and studied the course material will pass with a probability of 93%, and a deterministic naive Bayes classifier will yield "passed" for this student, and all other students with the same feature values. Table 8.2 shows the classifier probabilities and the values of the classifier function for all four possible feature vectors.

The major advantages of the naive Bayes classifier are its efficiency, because the training data have to be evaluated only once (to compute the relative frequencies), and that missing data can be simply ignored. The major disadvantages are that the features must be stochastically independent, which is often not the case in real-world data, and that the features must be discrete. Continuous features may be used in the naive Bayes classifier after discretization, as described in Chaps. 4 (histograms) and 5 (chi-square test).

Table 8.2 Naive Bayes classifier: classifier function for the student example

| | | Probability for | | Classification |
		Passed	Failed	
Went to class	Studied material	93%	7%	Passed
Went to class	Did not study material	67%	33%	Passed
Did not go to class	Studied material	46%	54%	Failed
Did not go to class	Did not study material	11%	89%	Failed

8.3   Linear Discriminant Analysis

The left view of Fig. 8.3 shows a scatter plot of a two-dimensional real-valued data set with two classes. The feature vectors of class 1 and 2 are shown as crosses and circles, respectively. Most points above the dotted line belong to class one (crosses), and most points below the dotted line belong to class two (circles). So the dotted line can be used as a class border, a so-called discriminant line, written in normal form as

$$w \cdot x^T + b = 0, \quad w \in \mathbb{R}^p, \quad b \in \mathbb{R} \tag{8.21}$$

For higher dimensions this denotes a discriminant plane ($p = 3$) or a discriminant hyperplane ($p > 3$). Without loss of generality we illustrate only two-dimensional linear discrimination here ($p = 2$). For a given data set Z, linear discriminant analysis finds a discriminant line (plane or hyperplane) by finding the parameters w and b that are optimal with respect to some well-defined criterion. For simplicity we restrict to two classes here. Multiple classes can be realized by multiple discriminant lines (planes or hyperplanes) or by combining several two-class classifiers.

The data in the left view of Fig. 8.3 are approximately Gaussian with means μ_1, μ_2 and approximately equal standard deviations $\sigma_1 \approx \sigma_2$. In this case, for symmetry reasons a suitable discriminant line is the perpendicular bisector of the line joining the two means, so the parameters of the discriminant line are

$$w = \mu_1 - \mu_2 \tag{8.22}$$

$$b = -w \cdot \frac{\mu_1^T + \mu_2^T}{2} \tag{8.23}$$

The dotted line in the left view of Fig. 8.3 is computed using these equations.

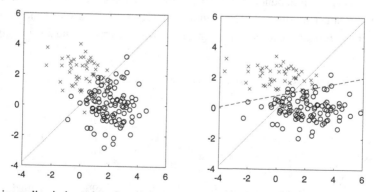

Fig. 8.3 Linear discriminant based on the perpendicular bisector of the line joining the means (left, dotted line) and linear discriminant analysis (right, dashed line)

The right view of Fig. 8.3 shows two classes with the same means μ_1 and μ_2 as in the left view, but the horizontal standard deviations are larger than the vertical standard deviations. The dotted line in the right view of Fig. 8.3 computed by (8.22) and (8.23) does not appropriately separate the classes in this case. *Linear discriminant analysis* is related to principal component analysis (Chap. 4). If we map the data to an orthogonal to the dotted line, then we obtain one-dimensional projections where the classes are well separated for the data in Fig. 8.3 (left), and not well separated for the data in Fig. 8.3 (right). Well separable means that in the projection the within-class variance

$$v_w = \sum_{i=1}^{c} \sum_{y_k=i} (x_k - \mu_i)^T (x_k - \mu_i) \tag{8.24}$$

is low, and the between-class variance

$$v_b = \sum_{i=1}^{c} (\mu_i - \bar{x})^T (\mu_i - \bar{x}) \tag{8.25}$$

is high. Therefore, Fisher's linear discriminant analysis [10] (that was originally used to analyze the Iris data set, see Chap. 2) maximizes the quotient

$$J = \frac{w^T \cdot v_b \cdot w}{w^T \cdot v_w \cdot w} \tag{8.26}$$

Similar to principal component analysis, this maximization yields the eigenproblem

$$(v_b^{-1} v_w) \cdot w = \lambda \cdot w \tag{8.27}$$

that can be solved for w, and then b can be computed by (8.23). For $c = 2$ we can simplify (8.25) to

$$v_b = (\mu_1 - \mu_2)^T (\mu_1 - \mu_2) \tag{8.28}$$

and maximizing J yields

$$w = v_w^{-1} (\mu_1 - \mu_2) \tag{8.29}$$

and (8.23). The dashed linear discriminant line in Fig. 8.3 (right) is computed using (8.29) and (8.23). It separates the classes much better than the dotted line.

The major advantages of linear discriminant analysis are its efficiency and its suitability for correlated features. The major disadvantage is that it assumes an approximately Gaussian distribution of the features and is only suited for linear class borders.

8.4   Support Vector Machine

A *support vector machine (SVM)* [19] is also based on linear class borders, but with a margin around the discriminant hyperplane. For two classes, the SVM requires that

$$w \cdot x_k^T + b \geq +1 \quad \text{if } y_k = 1 \tag{8.30}$$

$$w \cdot x_k^T + b \leq -1 \quad \text{if } y_k = 2 \tag{8.31}$$

where $b > 0$ is the margin parameter. If these constraints possess multiple (possibly infinitely many) solutions, then the SVM finds the solution with minimal

$$J = \|w\|^2 \tag{8.32}$$

which corresponds to maximal margin b. For closely adjacent or even overlapping classes no class border can satisfy (8.30) and (8.31), so these conditions are relaxed to

$$w \cdot x_k^T + b \geq +1 - \xi_k \quad \text{if } y_k = 1 \tag{8.33}$$

$$w \cdot x_k^T + b \leq -1 + \xi_k \quad \text{if } y_k = 2 \tag{8.34}$$

with non-zero slack variables $\xi_1, \ldots, \xi_n$ for each of the data points violating (8.30) or (8.31). To keep the nonzero slack variables low, a penalty term is added to the cost function (8.32) which yields

$$J = \|w\|^2 + \gamma \cdot \sum_{k=1}^{n} \xi_k, \quad \gamma > 0 \tag{8.35}$$

The SVM parameters w and b can be found by minimizing (8.35) under the constraints (8.33) and (8.34). The usual method for solving this constrained optimization problem is quadratic programming. See the references in the Appendix for the details of this method.

A modification of this approach represents the normal vector of the discriminant hyperplane as a linear combination of the training data

$$w = \sum_{y_j=1} \alpha_j x_j - \sum_{y_j=2} \alpha_j x_j \tag{8.36}$$

This expression corresponds to LDA for

$$\alpha_j = 1/|\{y_k = y_j \mid k = 1, \ldots, n\}|, \quad j = 1, \ldots, n \tag{8.37}$$

For SVM, the weight parameters $\alpha_1, \ldots, \alpha_n$ are optimized in order to minimize J (8.35). This yields the classification constraints

$$\sum_{y_j=1} \alpha_j x_j x_k^T - \sum_{y_j=2} \alpha_j x_j x_k^T + b \geq +1 - \xi_k \quad \text{if } y_k = 1 \tag{8.38}$$

$$\sum_{y_j=1} \alpha_j x_j x_k^T - \sum_{y_j=2} \alpha_j x_j x_k^T + b \leq -1 + \xi_k \quad \text{if } y_k = 2 \tag{8.39}$$

This modification increases the number of free parameters from $p + 1$ to $n + 1$, which may make the optimization harder, but allows the application of the so-called *kernel trick* which extends the SVM to nonlinear class borders. The basic idea which underlies the kernel trick is to map a data set $X = \{x_1, \ldots, x_n\} \in \mathbb{R}^p$ to a higher-dimensional data set $X' = \{x'_1, \ldots, x'_n\} \in \mathbb{R}^q$, $q > p$, so that the structure of the data in X' is more suitable than the structure in X. For the SVM we use the kernel trick to map classes with nonlinear class borders in X to classes with (approximately) linear class borders in X' and then use the linear classification approach presented above. In Chap. 9 we will also present an application of the kernel trick to relational cluster analysis. The mathematical basis of the kernel trick is *Mercer's theorem* [14] that was first published more than a hundred years ago and has gained a lot of attention in recent years because of its application to the kernel trick. Mercer's theorem states that for any data set X and any kernel function $k : \mathbb{R}^p \times \mathbb{R}^p \to \mathbb{R}$ there exists a mapping $\varphi : \mathbb{R}^p \to \mathbb{R}^q$ so that

$$k(x_j, x_k) = \varphi(x_j) \cdot \varphi(x_k)^T \tag{8.40}$$

This means that a mapping from X to X' can be implicitly done by replacing scalar products in X' by kernel function values in X, without explicitly specifying φ, and without explicitly computing X'. Replacing scalar products in X' by kernel functions in X is called the *kernel trick*. Some frequently used kernel functions are

- linear kernel

$$k(x_j, x_k) = x_j \cdot x_k^T \tag{8.41}$$

- polynomial kernel

$$k(x_j, x_k) = (x_j \cdot x_k^T)^d, \quad d \in \{2, 3, \ldots\} \tag{8.42}$$

- Gaussian kernel

$$k(x_j, x_k) = e^{-\frac{\|x_j - x_k\|^2}{\sigma^2}}, \quad \sigma > 0 \tag{8.43}$$

• hyperbolic tangent kernel

$$k(x_j, x_k) = 1 - \tanh \frac{\|x_j - x_k\|^2}{\sigma^2}, \quad \sigma > 0 \tag{8.44}$$

• radial basis function (RBF) kernel [16]

$$k(x_j, x_k) = f\left(\|x_j - x_k\|\right) \tag{8.45}$$

The Gaussian and hyperbolic tangent kernels are special cases of the RBF kernel. We apply
the kernel trick to (8.38) and (8.39), and finally obtain the SVM classification constraints

$$\sum_{y_j=1} \alpha_j k(x_j, x_k) - \sum_{y_j=2} \alpha_j k(x_j, x_k) + b \geq +1 - \xi_k \quad \text{if } y_k = 1 \tag{8.46}$$

$$\sum_{y_j=1} \alpha_j k(x_j, x_k) - \sum_{y_j=2} \alpha_j k(x_j, x_k) + b \leq -1 + \xi_k \quad \text{if } y_k = 2 \tag{8.47}$$

Minimization of J (8.35) with respect to these constraints yields the parameters $\alpha_1, \ldots, \alpha_n$
and b for the SVM classification rules

$$\sum_{y_j=1} \alpha_j k(x_j, x) - \sum_{y_j=2} \alpha_j k(x_j, x) + b \geq 0 \quad \Rightarrow \quad y = 1 \tag{8.48}$$

$$\sum_{y_j=1} \alpha_j k(x_j, x) - \sum_{y_j=2} \alpha_j k(x_j, x) + b \leq 0 \quad \Rightarrow \quad y = 2 \tag{8.49}$$

The major advantage of the SVM is that it can find nonlinear class borders. The major
disadvantage is that it requires to solve a very high-dimensional optimization problem
with nonlinear constraints that is computationally relatively expensive.

8.5   Nearest Neighbor Classifier

A much simpler classifier method is the *nearest neighbor classifier* [20] which assigns an
object with a given feature vector to the class of the training object with the most similar
feature vector. More formally, for a given feature vector x the nearest neighbor classifier
yields class y_k if

$$\|x - x_k\| = \min_{j=1,\ldots,n} \|x - x_j\| \tag{8.50}$$

where $\|.\|$ is a suitable dissimilarity measure, for example the Euclidean or the Maha-
lanobis distance. In case of multiple minima one of the minima may be chosen randomly.

Fig. 8.4 Class border
(Voronoi diagram) for a nearest
neighbor classifier

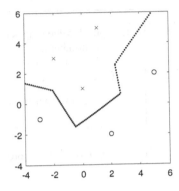

The resulting class border is piecewise linear along edges of a so-called *Voronoi diagram* like the one shown in Fig. 8.4. For noisy data or overlapping classes the nearest neighbor classifier sometimes yields bad results near the noise data or the class borders. This can be avoided by the *k-nearest neighbor classifier* which not only considers the nearest neighbor but the $k \in \{2, \ldots, n\}$ nearest neighbors and yields the most frequent class of these neighbors. For two classes, ties can be avoided by selecting odd values for k. Don't confuse the number k of nearest neighbors with the object index k.

The nearest neighbor and k-nearest neighbor classifiers do not explicitly design a classifier function but simply store the training data and only evaluate the data on a classification request. This behavior is called *lazy learning* [1]. The major advantages of the (k-)nearest neighbor classifier is that no explicit learning is required, that new training data can be easily included, and that nonlinear class borders can be identified. The major disadvantage is the high computational complexity of the classification, because for each classification the dissimilarities with each training feature vector have to be computed.

8.6   Learning Vector Quantization

Many approaches have been suggested to reduce the computational complexity of the nearest neighbor classification, for example by extracting a number of representative feature vectors (so-called *prototypes*) from the training data set and only computing the dissimilarities with these prototypes. A popular example for these approaches is *learning vector quantization (LVQ)* [12, 13], a heuristic algorithm that finds one prototype $v_i \in \mathbb{R}^p$, $i = 1, \ldots, c$, for each class and then assigns a feature vector x to class i if

$$\|x - v_i\| = \min_{j=1,\ldots,c} \|x - v_j\| \tag{8.51}$$

In LVQ training the prototypes are first randomly initialized. For each training vector the nearest prototype is moved towards the training vector if it belongs to the same class, and away from the training vector if it belongs to a different class. Notice that this scheme

1. input: labeled data set $X = \{x_1, \ldots, x_n\} \subset \mathbb{R}^p$, $y = \{y_1, \ldots, y_n\} \subset \{1, \ldots, c\}$, class number $c \in \{2, \ldots, n-1\}$, step length $\alpha(t)$
2. initialize $V = \{v_1, \ldots, v_c\} \subset \mathbb{R}^p$, $t = 1$
3. for $k = 1, \ldots, n$

    a. find winner prototype $v_i \in V$ with

$$\|x_k - v_i\| \leq \|x_k - v\| \quad \forall v \in V$$

    b. move winner prototype

$$v_i = \begin{cases} v_i + \alpha(t)(x_k - v_i) & \text{if } y_k = i \\ v_i - \alpha(t)(x_k - v_i) & \text{otherwise} \end{cases}$$

4. $t = t + 1$
5. repeat from (3.) until termination criterion holds
6. output: prototypes $V = \{v_1, \ldots, v_c\} \subset \mathbb{R}^p$

Fig. 8.5 Learning vector quantization (training)

is similar to the exponential filter presented in Chap. 3. To enforce convergence the step length is successively reduced during training. The algorithm passes through the training data set several times until a suitable termination criterion holds. Figure 8.5 presents the LVQ training algorithm in detail. While LVQ assigns each feature vector to one class, *fuzzy learning vector quantization (FLVQ)* [4, 7] partially assigns each vector to several classes.

The major advantage of LVQ compared to the (k-)nearest neighbor classifier is that it only has to find the minimum of c and not n dissimilarities for each classification, so it is computationally much cheaper for $c \ll n$. The major disadvantages are that it requires a training phase where each training vector has to be processed several times and that it can only realize quite simple class structures where each class is represented by only one prototype. For more complicated nonlinear class structures, various LVQ extensions have been proposed that use multiple prototypes per class. For an overview see [5].

8.7 Decision Trees

All previously presented classifiers consider all p features at a time. This may be disadvantageous if not all features are necessary for a good classification, and if the assessment of the feature values causes a high effort. In medical diagnosis, for example, body temperature or blood pressure are measured for many patients because these features are assessed with low effort and yield significant information about the patient's health, whereas more expensive or time-consuming tests are only performed if needed to support a specific diagnosis. Depending on the already observed feature data, additional features

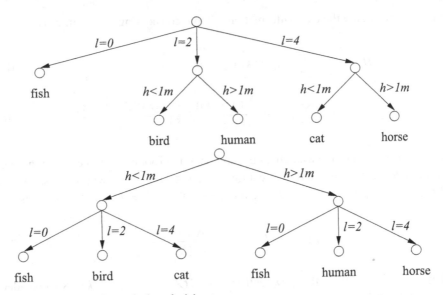

Fig. 8.6 Two functionally equivalent decision trees

may be ranked by importance, and only a subset of all possible features is used. This scheme leads to a *hierarchical* structure of the features, as opposed to the flat feature structure of the classifiers presented so far. Hierarchical classifiers can be represented by *decision trees* [18]. Figure 8.6 shows two different decision trees that realize the same classifier function. Both classifiers use the two-dimensional feature vector $x = (l, h)$ of a creature, where $l \in \{0, 2, 4\}$ is the number of legs, and $h > 0$ is the body height (in meters). Based on these two features the creature is classified as fish, bird, human, cat, or horse, so five classes are considered here. The first decision tree (top view of Fig. 8.6) considers first the feature l. For $l = 0$ the classification is terminated, the feature h can be ignored, and the result is fish. For $l = 2$ and $l = 4$ the set of possible classes is reduced to bird or human ($l = 2$) or cat or horse ($l = 4$), so the classification is not terminated, but the first feature yielded an *information gain*, and the class is finally determined by considering the second feature h. The second decision tree (bottom view of Fig. 8.6) considers first feature h, yields an information gain, and finalizes the classification by considering the second feature l.

Both decision trees realize the same classifier function but they yield different information gains in each step, depending on the order in which the features are considered. Given the training data $Z = (X, Y)$, the probability for class k is

$$p(y = k) = \frac{|\{y \in Y \mid y = k\}|}{|Y|} \tag{8.52}$$

so the entropy of only the class information without considering any features is

$$H(Z) = -\sum_{k=1}^{c} p(y = k) \log_2 p(y = k) \tag{8.53}$$

$$= -\sum_{k=1}^{c} \frac{|\{y \in Y \mid y = k\}|}{|Y|} \log_2 \frac{|\{y \in Y \mid y = k\}|}{|Y|} \tag{8.54}$$

This entropy quantifies the available information at the root node of the decision tree. At the root node we choose one of the features $x^{(j)}$, $j \in \{1, \ldots, p\}$, with the discrete range $x^{(j)} \in \{1, \ldots, v_j\}$, $v_j \in \{1, 2, \ldots\}$. The probability that $x^{(j)} = k$, $k = 1, \ldots, v_j$, is

$$p(x^{(j)} = k) = \frac{|\{x \in X \mid x^{(j)} = k\}|}{|X|} \tag{8.55}$$

and if $x^{(j)} = k$, then the new entropy is $H(Z \mid x^{(j)} = k)$. So considering feature $x^{(j)}$ yields an expected information gain of $H(Z)$ minus the expected value of $H(Z \mid x^{(j)} = k)$.

$$g_j = H(Z) - \sum_{k=1}^{v_j} p(x^{(j)} = k) \cdot H(Z \mid x^{(j)} = k) \tag{8.56}$$

$$= H(Z) - \sum_{k=1}^{v_j} \frac{|\{x \in X \mid x^{(j)} = k\}|}{|X|} H(Z \mid x^{(j)} = k) \tag{8.57}$$

In a recursive manner, at each node from the list of remaining features we pick the one with the highest expected information gain. For our example in Fig. 8.6 consider the data sets X and Y given in Table 8.3. The data set contains one fish, one bird, one horse, three humans, and two cats, so the entropy is

$$H(Z) = -\frac{1}{8} \log_2 \frac{1}{8} - \frac{1}{8} \log_2 \frac{1}{8} - \frac{3}{8} \log_2 \frac{3}{8} - \frac{2}{8} \log_2 \frac{2}{8} - \frac{1}{8} \log_2 \frac{1}{8}$$

$$\approx 2.1556 \text{ bit} \tag{8.58}$$

Table 8.3 Data set for Fig. 8.6

Index	1	2	3	4	5	6	7	8
Height h [m]	0.1	0.2	1.8	0.2	2.1	1.7	0.1	1.6
Legs l	0	2	2	4	4	2	4	2
Class	Fish	Bird	Human	Cat	Horse	Human	Cat	Human

We first consider the decision tree in the top view of Fig. 8.6, where the feature $l \in \{0, 2, 4\}$ is considered first. The probabilities of the feature values of l are

$$p(l = 0) = \frac{1}{8} \tag{8.59}$$

$$p(l = 2) = \frac{4}{8} \tag{8.60}$$

$$p(l = 4) = \frac{3}{8} \tag{8.61}$$

and the conditional entropies are

$$H(Z \mid l = 0) = -1 \log_2 1 = 0 \tag{8.62}$$

$$H(Z \mid l = 2) = -\frac{1}{4} \log_2 \frac{1}{4} - \frac{3}{4} \log_2 \frac{3}{4} \approx 0.8113 \text{ bit} \tag{8.63}$$

$$H(Z \mid l = 4) = -\frac{1}{3} \log_2 \frac{1}{3} - \frac{2}{3} \log_2 \frac{2}{3} \approx 0.9183 \text{ bit} \tag{8.64}$$

So at the root node the expected information gain for feature l is

$$g_l = H(Z) - \frac{1}{8} H(Z \mid l = 0) - \frac{4}{8} H(Z \mid l = 2) - \frac{3}{8} H(Z \mid l = 4)$$
$$\approx 1.4056 \text{ bit} \tag{8.65}$$

Next we consider the decision tree in the bottom view of Fig. 8.6, where the feature $h \in \{< 1m, > 1m\}$ is considered first. The probabilities of the feature values of h are

$$p(h < 1m) = \frac{4}{8} \tag{8.66}$$

$$p(h > 1m) = \frac{4}{8} \tag{8.67}$$

and the conditional entropies are

$$H(Z \mid h < 1m) = -\frac{1}{4} \log_2 \frac{1}{4} - \frac{1}{4} \log_2 \frac{1}{4} - \frac{2}{4} \log_2 \frac{2}{4} \approx 1.5 \text{ bit} \tag{8.68}$$

$$H(Z \mid h > 1m) = -\frac{1}{4} \log_2 \frac{1}{4} - \frac{3}{4} \log_2 \frac{3}{4} \approx 0.8113 \text{ bit} \tag{8.69}$$

1. input: $X = \{x_1, \ldots, x_n\} \subset \{1, \ldots, v_1\} \times \ldots \times \{1, \ldots, v_p\}$, $Y = \{y_1, \ldots, y_n\} \subset \{1, \ldots, c\}$
2. create root node W
3. ID3$(X, Y, W, \{1, \ldots, p\})$
4. output: tree with root W

procedure ID3(X, Y, N, I)

1. if I is empty or all Y are equal then terminate
2. compute information gain $g_i(X, Y)\ \forall i \in I$
3. determine winner feature $j = \operatorname{argmax}\{g_i(X, Y)\}$
4. partition X, Y into v_j disjoint subsets

$$X_i = \{x_k \in X \mid x_k^{(j)} = i\}, \quad Y_i = \{y_k \in Y \mid x_k^{(j)} = i\}, \quad i = 1, \ldots, v_j$$

5. for i with $X_i \neq \{\}, Y_i \neq \{\}$

   a. create new node N_i and append it to N
   b. ID3$(X_i, Y_i, N_i, I \setminus \{j\})$

Fig. 8.7 ID3 algorithm

So at the root node the expected information gain for feature h is

$$g_h = H(Z) - \frac{4}{8}H(Z \mid h < 1m) - \frac{4}{8}H(Z \mid h > 1m)$$

$$\approx 1 \text{ bit} \tag{8.70}$$

At the root node, the information gain of feature l is higher than the information gain of feature h, $g_l > g_h$, so in our example the optimal decision tree is the one shown in the *top* view of Fig. 8.6 that first considers l then h.

Optimal decision trees can be constructed using the *iterative dichotomiser 3 (ID3) algorithm* [17] shown in Fig. 8.7. The recursive procedure is called with the data sets X and Y, the root of the decision tree to be constructed, and the set of all feature indices. The recursion stops if all data belong to the same class, because then no further branches are needed. Otherwise the algorithm computes the information gain for the remaining features and determines the winner feature. The data are partitioned according to the values of the winner feature, and for each nonempty subset a new node is created and appended. For each appended node the algorithm recursively computes the corresponding sub-tree.

ID3 requires discrete features. An extension of ID3 to real-valued features is the *classification and regression tree (CART)* algorithm [6] that finds optimal interval borders (like $< 1m$ and $> 1m$ in our example) that maximize the entropy. The *chi-square automatic interaction detection (CHAID)* algorithm [11] finds the interval borders using a chi-squared test. Other extensions of ID3 such as *C4.5* and *C5.0* accept real-valued and missing features, use *pruning* mechanisms to reduce the tree size and enable exportation of the decision tree in form of decision *rules*.

The major advantages of decision tree classifiers are that they do not necessarily consider all features, which is useful for data with a large number of features, and that they provide additional information about feature ranks. The major disadvantage is that they are either restricted to discrete features, or they work with continuous features but the class borders can only be (at least piecewise) parallel to the coordinate axes, so complicated nonlinear class borders cannot be efficiently modeled.

Problems

8.1 Consider the data sets for two classes $X_1 = \{(0, 0)\}$ and $X_2 = \{(1, 0), (0, 1)\}$.

(a) Which classification probabilities will a naive Bayes classifier produce for the feature vector $(0, 0)$?
(b) Which discriminant line will be found by a support vector machine with linear kernels and infinite penalty for margin violations?
(c) Which classification rule will be found by a nearest neighbor classifier?
(d) Which classification rule will be found by a three nearest neighbor classifier?
(e) Draw a receiver operating characteristic for these four classifiers, where class 2 is considered positive.

8.2 Consider a classifier with one-dimensional input $x \in \mathbb{R}$, where the data for the positive and the negative class follow the Cauchy distributions

$$p(x \mid \oplus) = \frac{1}{\pi} \cdot \frac{1}{1 + (x + 1)^2} \quad \text{and} \quad p(x \mid \ominus) = \frac{1}{\pi} \cdot \frac{1}{1 + (x - 1)^2}$$

(a) For 50% positive and 50% negative training data, for which x will the naive Bayes classifier yield "positive"?
(b) For which percentage p of positive training data will the naive Bayes classifier *always* yield "positive"? If needed, assume $\sqrt{2} \approx 1.4$.
(c) For an arbitrary percentage P of positive training data, for which x will the naive Bayes classifier yield "positive"?

8.3 We build a support vector machine classifier for the XOR data set defined by $X_- = \{(-1, -1), (1, 1)\}$ and $X_+ = \{(-1, 1), (1, -1)\}$.

(a) Are the classes of the XOR data set linearly separable?
(b) We use the quadratic kernel $k(x, y) = (xy^T)^2$ and explicitly map the data to $\mathbb{R}^3$, where we use the squares of the original two-dimensional data as the first two dimensions. Determine the formula to compute the third dimension.

(c) In the three-dimensional space from (b), determine the normal vector, offset, and margin of the separation plane that the support vector machine will find.

8.4 A car manufacturer has produced 50.000 cars with 90 kW gasoline engine, 100.000 cars with 120 kW gasoline engine, and 50.000 cars with 90 kW diesel engine. For all calculations use the following assumptions: $-\frac{1}{4}\log_2\frac{1}{4}=\frac{1}{2}$, $-\frac{1}{3}\log_2\frac{1}{3}=\frac{19}{36}$, $-\frac{1}{2}\log_2\frac{1}{2}=\frac{1}{2}$, $-\frac{2}{3}\log_2\frac{2}{3}=\frac{14}{36}$.

(a) What is the entropy of these data?
(b) For a given car, how much information do you gain if you know the kW value?
(c) For a given car, how much information do you gain if you know whether it has a gasoline or diesel engine?
(d) Sketch the ID3 decision tree.

References

1. D. W. Aha. Editorial: Lazy learning. *Artificial Intelligence Review (Special Issue on Lazy Learning)*, 11(1–5):7–10, June 1997.
2. R. Baeza-Yates and B. Ribeiro-Neto. *Modern Information Retrieval*. Addison-Wesley, New York, 1999.
3. T. Bayes. An essay towards solving a problem in the doctrine of chances. *Philosophical Transactions of the Royal Society of London*, 53:370–418, 1763.
4. J. C. Bezdek and N. R. Pal. Two soft relatives of learning vector quantization. *Neural Networks*, 8(5):729–743, 1995.
5. J. C. Bezdek, T. R. Reichherzer, G. S. Lim, and Y. Attikiouzel. Multiple-prototype classifier design. *IEEE Transactions on Systems, Man, and Cybernetics C*, 28(1):67–79, 1998.
6. L. Breiman, J. H. Friedman, R. A. Olsen, and C. J. Stone. *Classification and Regression Trees*. Chapman & Hall, New Work, 1984.
7. F. L. Chung and T. Lee. Fuzzy learning vector quantization. In *IEEE International Joint Conference on Neural Networks*, volume 3, pages 2739–2743, Nagoya, October 1993.
8. J. Davis and M. Goadrich. The relationship between precision–recall and ROC curves. In *International Conference on Machine Learning*, pages 233–240, 2006.
9. R. O. Duda and P. E. Hart. *Pattern Classification and Scene Analysis*. Wiley, New York, 1973.
10. R. A. Fisher. The use of multiple measurements in taxonomic problems. *Annals of Eugenics*, 7:179–188, 1936.
11. G. V. Kass. Significance testing in automatic interaction detection (AID). *Applied Statistics*, 24:178–189, 1975.
12. T. Kohonen. Learning vector quantization. *Neural Networks*, 1:303, 1988.
13. T. Kohonen. Improved versions of learning vector quantization. In *International Joint Conference on Neural Networks*, volume 1, pages 545–550, San Diego, June 1990.
14. J. Mercer. Functions of positive and negative type and their connection with the theory of integral equations. *Philosophical Transactions of the Royal Society A*, 209:415–446, 1909.
15. J. Neyman and E. S. Pearson. Interpretation of certain test criteria for purposes of statistical inference, part I. *Joint Statistical Papers, Cambridge University Press*, pages 1–66, 1967.

16. M. J. D. Powell. Radial basis functions for multi–variable interpolation: a review. In *IMA Conference on Algorithms for Approximation of Functions and Data*, pages 143–167, Shrivenham, 1985.

17. J. R. Quinlan. Induction on decision trees. *Machine Learning*, 11:81–106, 1986.

18. L. Rokach and O. Maimon. *Data Mining with Decision Trees: Theory and Applications*. Machine Perception and Artificial Intelligence. World Scientific Publishing Company, 2008.

19. B. Schölkopf and A. J. Smola. *Learning with Kernels*. MIT Press, Cambridge, 2002.

20. G. Shakhnarovich, T. Darrell, and P. Indyk. *Nearest–Neighbor Methods in Learning and Vision: Theory and Practice*. Neural Information Processing. MIT Press, 2006.

21. S. Theodoridis and K. Koutroumbas. *Pattern Recognition*. Academic Press, San Diego, 4th edition, 2008.

Clustering

9

Abstract

Clustering is unsupervised learning that assigns labels to objects in unlabeled data. When clustering is performed on data that possess class labels, the clusters may or may not correspond with these classes. Cluster partitions may be mathematically represented by sets, partition matrices, and/or cluster prototypes. Sequential clustering (single linkage, complete linkage, average linkage, Ward's method, etc.) yields hierarchical cluster structures but is computationally expensive. Partitional clustering can be based on hard, fuzzy, possibilistic, or noise clustering models. Cluster prototypes can have different shapes such as hyperspheres, ellipsoids, lines, circles, or more complex shapes. Relational clustering finds clusters in relational data, often enhanced by kernelization. Cluster tendency assessment finds out if the data contain clusters at all, and cluster validity measures help identify the number of clusters or other algorithmic parameters. Clustering can also be done by heuristic methods such as self-organizing maps.

9.1 Cluster Partitions

The previous chapter on classification considered data sets that contain feature vectors X and class labels Y. In many applications class labels are not available or difficult to obtain, for example by manual labeling. *Clustering* [18, 19] identifies structures in *unlabeled* feature data X. When clustering is performed on data with class labels (such as the Iris data), the clusters may or may not correspond with these classes, however, clusters are often used as class estimates. Figure 9.1 illustrates the clustering methods presented in this book. We distinguish sequential and prototype-based clustering. Prototype-based clustering may use different clustering models (hard, fuzzy, possibilistic, noise), prototypes

© Springer Fachmedien Wiesbaden GmbH, part of Springer Nature 2020

T. A. Runkler, *Data Analytics*, https://doi.org/10.1007/978-3-658-29779-4_9

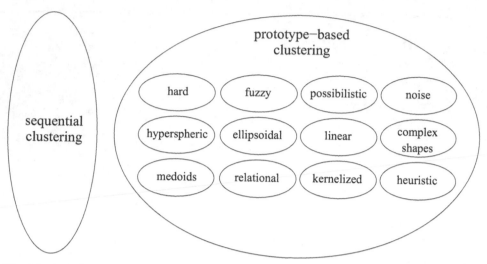

Fig. 9.1 Some important clustering methods

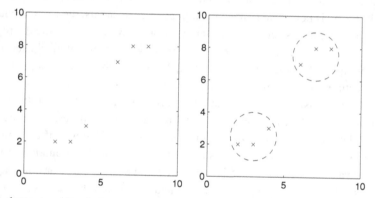

Fig. 9.2 A data set and its cluster structure

(hyperspheric, ellipsoidal, linear, complex shapes), and other characteristics (medoids, relational, kernelized, heuristic).

Figure 9.2 (left) shows a scatter plot of the data set

$$X = \{(2, 2), (3, 2), (4, 3), (6, 7), (7, 8), (8, 8)\} \tag{9.1}$$

that arguably contains two clusters, bottom left and top right, whose boundaries are illustrated by the dashed circles in Fig. 9.2 (right). The cluster structure partitions this data set X into the disjoint subsets $C_1 = \{x_1, x_2, x_3\}$ and $C_2 = \{x_4, x_5, x_6\}$. More generally, clustering partitions a data set $X = \{x_1, \ldots, x_n\} \subset \mathbb{R}^p$ into $c \in \{2, 3, \ldots, n - 1\}$ disjoint subsets $C_1, \ldots, C_c$ so that

$$X = C_1 \cup \ldots \cup C_c \tag{9.2}$$

$$C_i \neq \{\} \quad \text{for all } i = 1, \ldots, c \tag{9.3}$$

$$C_i \cap C_j = \{\} \quad \text{for all } i, j = 1, \ldots, c, i \neq j \tag{9.4}$$

9.2 Sequential Clustering

A popular family of *sequential clustering* methods is *sequential agglomerative hierarchical nonoverlapping (SAHN)* clustering [30] that is illustrated in Fig. 9.3. Initially, SAHN interprets each of the n objects as an individual cluster, so the initial partition is $\Gamma_n = \{\{o_1\}, \ldots, \{o_n\}\}$. In each step, SAHN finds the pair of least dissimilar (or most similar) clusters and merges these two clusters to a new cluster, so the number of clusters is decreased by one. SAHN terminates when the number of desired clusters is reached, or when all objects are merged to a single cluster which corresponds to the partition $\Gamma_1 = \{\{o_1, \ldots, o_n\}\}$. The dissimilarity between a pair of clusters can be computed from the dissimilarities between pairs of objects in these clusters in various ways:

- minimum distance or *single linkage*

$$d(C_r, C_s) = \min_{x \in C_r, \, y \in C_s} d(x, y) \tag{9.5}$$

- maximum distance or *complete linkage*

$$d(C_r, C_s) = \max_{x \in C_r, \, y \in C_s} d(x, y) \tag{9.6}$$

Fig. 9.3 Sequential agglomerative hierarchical nonoverlapping (SAHN) clustering

1. input: $X = \{x_1, \ldots, x_n\} \subset \mathbb{R}^p$
2. initialize $\Gamma_n = \{\{x_1\}, \ldots, \{x_n\}\}$
3. for $c = n - 1, n - 2, \ldots, 1$

   a. $(i, j) = \operatorname*{argmin}_{C_r, C_s \in \Gamma_c} d(C_r, C_s)$

   b. $\Gamma_c = (\Gamma_{c+1} \setminus C_i \setminus C_j) \cup \{C_i \cup C_j\}$

4. output: partitions $\Gamma_1, \ldots, \Gamma_n$

- average distance or *average linkage*

$$d(C_r, C_s) = \frac{1}{|C_r| \cdot |C_s|} \sum_{x \in C_r,\ y \in C_s} d(x, y) \tag{9.7}$$

These three options apply to relational or feature data. Two more options which are defined only for feature data are

- distance of the centers

$$d(C_r, C_s) = \left\| \frac{1}{|C_r|} \sum_{x \in C_r} x - \frac{1}{|C_s|} \sum_{x \in C_s} x \right\| \tag{9.8}$$

- Ward's measure [31]

$$d(C_r, C_s) = \frac{|C_r| \cdot |C_s|}{|C_r| + |C_s|} \left\| \frac{1}{|C_r|} \sum_{x \in C_r} x - \frac{1}{|C_s|} \sum_{x \in C_s} x \right\| \tag{9.9}$$

Figure 9.4 visualizes the first three options (single, complete, and average linkage). The left diagram shows the minimum and maximum distances (single and complete linkage) between the clusters from Fig. 9.2. The right diagram shows all mutual distances that are used in average linkage. Single linkage is one of the most popular cluster distance measures. However, single linkage tends to form long stringy clusters. SAHN algorithms with specific cluster distance measures are often denoted by the measure only, for example SAHN with the single linkage distance is simply called *single linkage clustering*.

The partitions produced by SAHN represent hierarchical structures that can be visualized by a so-called *dendrograms*. Figure 9.5 shows the single linkage dendrogram for our simple six point data set, that illustrates how the points $x_1, \ldots, x_6$ (indices on the

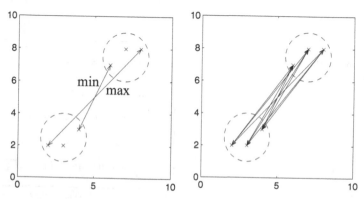

Fig. 9.4 Distances between clusters (left: single and complete linkage, right: average linkage)

Fig. 9.5 Dendrogram

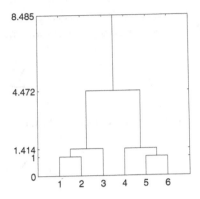

horizontal axis) are successively merged in a bottom-up scheme. The vertical axis shows the single linkage distances. The single linkage partitions are

$$\Gamma_6 = \{\{x_1\}, \{x_2\}, \{x_3\}, \{x_4\}, \{x_5\}, \{x_6\}\} \tag{9.10}$$

$$\Gamma_4 = \Gamma_5 = \{\{x_1, x_2\}, \{x_3\}, \{x_4\}, \{x_5, x_6\}\} \tag{9.11}$$

$$\Gamma_2 = \Gamma_3 = \{\{x_1, x_2, x_3\}, \{x_4, x_5, x_6\}\} \tag{9.12}$$

$$\Gamma_1 = = \{\{x_1, x_2, x_3, x_4, x_5, x_6\}\} = \{X\} \tag{9.13}$$

A variant of SAHN uses *density estimators* instead of distance measures to merge clusters: *density-based spatial clustering of applications with noise (DBSCAN)* [28].

The reverse or top-down variant of SAHN is *sequential divisive hierarchical nonoverlapping (SDHN)* clustering which starts with all points in one cluster and successively separates clusters into sub-clusters, so $c_0^* = 1$ and $\Gamma_0^* = \{\{x_1, \ldots, x_n\}\}$. Separating clusters is not a trivial task, so SDHN is less popular than SAHN. An example of literature on SDHN is [11].

The main advantages of SAHN are that it yields hierarchical cluster structures. The main disadvantage is the high computational complexity $(o(n^3)$ for the algorithm in Fig. 9.3). Notice that more efficient algorithms for SAHN $(o(n^2 \log n))$ and single linkage $(o(n^2))$ have been presented in the literature [10].

9.3 Prototype-Based Clustering

In the previous section we represented cluster structures of a data set X by a *partition set* Γ containing disjoint subsets of X. An equivalent representation is a *partition matrix U* with the elements

$$u_{ik} = \begin{cases} 1 & \text{if } x_k \in C_i \\ 0 & \text{if } x_k \notin C_i \end{cases} \tag{9.14}$$

$i = 1, \ldots, c, k = 1, \ldots, n$, so u_{ik} is a *membership value* that indicates if x_k belongs to C_i. For non-empty clusters we require

$$\sum_{k=1}^{n} u_{ik} > 0, \quad i = 1, \ldots, c \tag{9.15}$$

and for pairwise disjoint clusters

$$\sum_{i=1}^{c} u_{ik} = 1, \quad k = 1, \ldots, n \tag{9.16}$$

Figure 9.6 (left) shows our six point data set again with vertical bars representing one row of the partition matrix. The elements of this row of the partition matrix are one for points in the corresponding cluster and zero for all other points. Therefore, there are three bars with height one at the points $x_4, \ldots, x_6$, and there are no bars (zero height) at the other points.

Besides partition sets and partition matrices also cluster *prototypes* can be used to represent clusters when the data are feature vectors. For example, each cluster may be represented by a (single) center v_i, $i = 1, \ldots, c$, so the cluster structure is defined by the set of cluster centers

$$V = \{v_1, \ldots, v_c\} \subset \mathbb{R}^p \tag{9.17}$$

Given a data set X, the cluster centers V and the partition matrix U can be found by optimizing the *c-means (CM)* clustering model [1] defined as the sum of the square distances between the cluster centers and the data points belonging to the respective clusters.

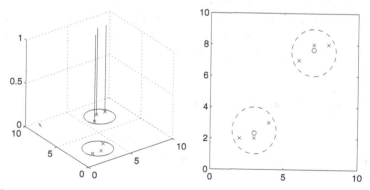

Fig. 9.6 Cluster memberships and cluster centers

$$J_{CM}(U, V; X) = \sum_{i=1}^{c} \sum_{x_k \in C_i} \|x_k - v_i\|^2 = \sum_{i=1}^{c} \sum_{k=1}^{n} u_{ik} \|x_k - v_i\|^2 \qquad (9.18)$$

To minimize J_{CM}, for each k we pick the minimum $\|x_k - v_i\|$ and set the corresponding membership u_{ik} to one. This yields the partition matrix $U \in [0, 1]^{c \times p}$ with

$$u_{ik} = \begin{cases} 1 & \text{if } \|x_k - v_i\| = \min_{j=1,\dots,c} \|x_k - v_j\| \\ 0 & \text{otherwise} \end{cases} \qquad (9.19)$$

where ties have to be handled in an appropriate way, for example by random assignment. The necessary condition for extrema of (9.18)

$$\frac{\partial J_{CM}(U, V; X)}{\partial v_i} = 0, \quad i = 1, \dots, c \qquad (9.20)$$

yields the cluster centers

$$v_i = \frac{1}{|C_i|} \sum_{x_k \in C_i} x_k = \frac{\sum_{k=1}^{n} u_{ik} x_k}{\sum_{k=1}^{n} u_{ik}} \qquad (9.21)$$

So the cluster centers are the average of the data points assigned to the corresponding cluster, and the data points are assigned to clusters using the *nearest neighbor* rule.

Notice that the optimal V depends on U (and X), and the optimal U depends on V (and X), so this requires an *alternating optimization (AO)* of U and V, as shown in Fig. 9.7. This AO variant initializes V, alternatingly computes U and V, and terminates on V. A dual AO variant initializes U, alternatingly computes V and U, and terminates on U. The size of V is $c \times p$, and the size of U is $c \times n$, so for $p < n$, the AO variant shown in Fig. 9.7 is more efficient, and for $p > n$ the dual variant is more efficient.

9.4 Fuzzy Clustering

CM clustering works well if the clusters are well separated and do not contain inliers nor outliers. We add the point $x_7 = (5, 5)$ to the previously considered six point data set and obtain the new seven point data set

$$X = \{(2, 2), (3, 2), (4, 3), (6, 7), (7, 8), (8, 8), (5, 5)\} \qquad (9.22)$$

1. input: data $X = \{x_1, \ldots, x_n\} \subset \mathbb{R}^p$,
   cluster number $c \in \{2, \ldots, n-1\}$,
   maximum number of steps t_{max},
   distance measure $\|.\|$,
   distance measure for termination $\|.\|_\varepsilon$,
   termination threshold ε
2. initialize prototypes $V^{(0)} \subset \mathbb{R}^p$
3. for $t = 1, \ldots, t_{max}$

   a. compute $U^{(t)}(V^{(t-1)}, X)$
   b. compute $V^{(t)}(U^{(t)}, X)$
   c. if $\|V^{(t)} - V^{(t-1)}\|_\varepsilon \leq \varepsilon$ then stop

4. output: partition matrix $U \in [0,1]^{c \times n}$,
   prototypes $V = \{v_1, \ldots, v_c\} \in \mathbb{R}^p$

Fig. 9.7 Alternating optimization of clustering models

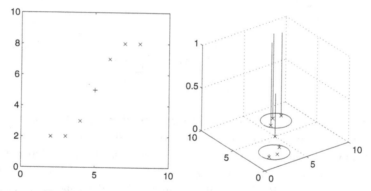

Fig. 9.8 Data set with a point that belongs to both clusters (left) and fuzzy partition (right)

Figure 9.8 (left) shows this data set, with the new data point shown as a plus sign. Assigning the additional data point to either of the two clusters would contradict intuition. Because the data set is symmetric about the new point, it seems reasonable to require partial membership in both clusters. A partial cluster assignment can be formally realized by allowing the elements of the partition matrix to be in the unit interval $u_{ik} \in [0, 1]$, $i = 1, \ldots, c, k = 1, \ldots, n$. In correspondence with (9.15) we require non-empty fuzzy clusters

$$\sum_{k=1}^{n} u_{ik} > 0, \quad i = 1, \ldots, c \qquad (9.23)$$

and in correspondence with (9.16) we require that for each point the cluster memberships sum up to one

$$\sum_{i=1}^{c} u_{ik} = 1, \quad k = 1, \ldots, n \tag{9.24}$$

Condition (9.24) is similar to the normalization of probabilistic distributions, but these set memberships are not probabilistic. For a discussion on the difference between fuzziness and probability we refer to [3]. Figure 9.8 (right) shows memberships of one row of a fuzzy partition. The memberships of the new point x_7 are $u_{17} = u_{27} = 1/2$. The memberships u_{21}, u_{22}, u_{23} are small but may be different from zero, and the memberships u_{24}, u_{25}, u_{26} are large but less than one to satisfy constraint (9.24). Each point is assigned to all (both) clusters to a certain degree.

Given a data set X, a set of prototypes V and a fuzzy partition matrix U can be found by optimizing the *fuzzy c-means (FCM)* clustering model [2, 7] defined by the objective function

$$J_{\text{FCM}}(U, V; X) = \sum_{k=1}^{n} \sum_{i=1}^{c} u_{ik}^{m} \|x_k - v_i\|^2 \tag{9.25}$$

with the constraints $0 \leq u_{ik} \leq 1$, (9.23) and (9.24). The parameter $m > 1$ specifies the *fuzziness* of the clusters. As $m \to 1$ from above, FCM becomes CM. For $m \to \infty$ each point in $X \setminus V$ will obtain the same membership $u_{ik} = 1/c, i = 1, \ldots, c, k = 1, \ldots, n$, in all clusters. A frequent choice is $m = 2$. In FCM, constraint (9.23) can be formally ignored but constraint (9.24) has to be explicitly considered in order to avoid the trivial unconstrained optimum $u_{ik} = 0, i = 1, \ldots, c, k = 1, \ldots, n$. Finding constrained extrema of J_{FCM} uses the Lagrange function (see Appendix)

$$L_{\text{FCM}}(U, V, \lambda; X) = \sum_{k=1}^{n} \sum_{i=1}^{c} u_{ik}^{m} \|x_k - v_i\|^2 - \sum_{k=1}^{n} \lambda_k \cdot \left(\sum_{i=1}^{c} u_{ik} - 1 \right) \tag{9.26}$$

The necessary conditions for extrema yield

$$\left. \begin{array}{c} \dfrac{\partial L_{\text{FCM}}}{\partial \lambda_k} = 0 \\[2ex] \dfrac{\partial L_{\text{FCM}}}{\partial u_{ik}} = 0 \end{array} \right\} \Rightarrow u_{ik} = 1 \left/ \sum_{j=1}^{c} \left(\dfrac{\|x_k - v_i\|}{\|x_k - v_j\|} \right)^{\frac{2}{m-1}} \right. \tag{9.27}$$

$$\frac{\partial J_{\text{FCM}}}{\partial v_i} = 0 \quad \Rightarrow \quad v_i = \frac{\sum_{k=1}^{n} u_{ik}^m x_k}{\sum_{k=1}^{n} u_{ik}^m} \tag{9.28}$$

Comparing (9.25) with (9.18), (9.27) with (9.19), and (9.28) with (9.21) confirms that as $m \to 1$ from above, FCM becomes CM. In FCM, U depends on V, and V depends on U, so U and V are found by alternating optimization (AO) as shown in Fig. 9.7. Other methods have been proposed to optimize FCM, for example evolutionary algorithms [6] or swarm intelligence [27].

FCM often yields good results, even if the clusters are overlapping and data are noisy, but it is sensitive to outliers. Outliers that are almost equally distant to all cluster centers receive the FCM memberships $u_{ik} \approx 1/c$, $i = 1, \ldots, c$, in order to satisfy the normalization criterion (9.24). So for FCM, outliers are equivalent to other data points that are equidistant to all data points like the middle point in Fig. 9.8. But intuitively we expect outliers to have low membership in all clusters because they are not representative for any cluster. This may be realized by dropping the normalization criterion (9.24) and adding a penalty term to the FCM objective function, which yields the *possibilistic c-means (PCM)* clustering model [23] with the objective function

$$J_{\text{PCM}}(U, V; X) = \sum_{k=1}^{n} \sum_{i=1}^{c} u_{ik}^m \|x_k - v_i\|^2 + \sum_{i=1}^{c} \eta_i \sum_{k=1}^{n} (1 - u_{ik})^m \tag{9.29}$$

with the constraint (9.23) and the cluster sizes $\eta_1, \ldots, \eta_c > 0$. The necessary conditions for extrema yield (9.28) and

$$u_{ik} = 1 \left/ \sum_{j=1}^{c} \left(1 + \left(\frac{\|x_k - v_i\|^2}{\eta_i} \right)^{\frac{1}{m-1}} \right) \right. \tag{9.30}$$

which is a so-called *Cauchy function*.

An alternative approach to handle outliers in fuzzy clustering defines an additional cluster specifically for outliers. The membership of each point in this outlier cluster is one minus the memberships in the regular clusters. The outlier cluster does not have a cluster center, but the distance between each data point and the outlier cluster is a constant $\delta > 0$ that has to be chosen higher than the distances between non-outliers and regular cluster centers. This yields the *noise clustering (NC)* model [9] with the objective function

$$J_{\text{NC}}(U, V; X) = \sum_{k=1}^{n} \sum_{i=1}^{c} u_{ik}^m \|x_k - v_i\|^2 + \sum_{k=1}^{n} \left(1 - \sum_{j=1}^{c} u_{jk} \right)^m \delta^2 \tag{9.31}$$

The necessary conditions for extrema yield (9.28) and

$$u_{ik} = 1 \left/ \left(\sum_{j=1}^{c} \left(\frac{\|x_k - v_i\|}{\|x_k - v_j\|} \right)^{\frac{2}{m-1}} + \left(\frac{\|x_k - v_i\|}{\delta} \right)^{\frac{2}{m-1}} \right) \right. \tag{9.32}$$

The distances in the objective functions of CM (9.18), FCM (9.25), PCM (9.29), NC (9.31) and in the corresponding equations for u_{ik} (9.19), (9.27), (9.30), (9.32) can be defined by any of the functions from Chap. 2, for example using the Euclidean norm or the Mahalanobis norm. A variant of the Mahalanobis norm considers the covariance matrices of the individual clusters instead of the whole data set. The covariance within cluster $i = 1, \ldots, c$ is defined as

$$S_i = \sum_{k=1}^{n} u_{ik}^m (x_k - v_i)^T (x_k - v_i) \tag{9.33}$$

which corresponds to the within-class covariance (8.24) for linear discriminant analysis (Chap. 8). The within-cluster covariance defines a matrix norm with

$$A_i = \sqrt[p]{\rho_i \det(S_i)} \, S_i^{-1} \tag{9.34}$$

which normalizes each cluster to the hypervolume $\det(A_i) = \rho_i$. A frequent choice is $\rho_1 = \ldots = \rho_c = 1$. FCM with the within-cluster Mahalanobis norm is also called the *Gustafson-Kessel (GK)* clustering model [13] with the objective function

$$J_{GK}(U, V; X) = \sum_{k=1}^{n} \sum_{i=1}^{c} u_{ik}^m \left((x_k - v_i) A_i (x_k - v_i)^T \right) \tag{9.35}$$

The within-cluster Mahalanobis norm can also be used in CM, PCM, and NC.

The Euclidean norm yields hyperspherical clusters, and the within-cluster Mahalanobis norm yields hyperellipsoidal clusters. Expanding our notion of prototypes to include structures more complex than points in the feature space allows us to model more complicated geometric cluster structures. A line, plane, or hyperplane can be defined by a point $v_i \in \mathbb{R}^p$ and one or more directions $d_{i1}, \ldots, d_{iq}, q \in \{1, \ldots, p-1\}$, where $q = 1$ for lines and $q = 2$ for planes. The distance of a point x_k and such a line, plane, or hyperplane follows from the orthogonal projection.

$$d(x_k, v_i, d_{i1}, \ldots, d_{iq}) = \sqrt{\|x_k - v_i\|^2 - \sum_{l=1}^{q} \left((x_k - v_i) d_{il}^T \right)^2} \tag{9.36}$$

Formally, the distance $\|x_k - v_i\|$ in any clustering model (CM, FCM, PCM, or NC) and the corresponding equations to compute U can be simply replaced by (9.36) to obtain a clustering model for lines, planes, or hyperplanes. For FCM, for example, we obtain the *fuzzy c-varieties (FCV)* clustering model [4]

$$J_{FCV}(U, V, D; X) = \sum_{k=1}^{n} \sum_{i=1}^{c} u_{ik}^m \left(\|x_k - v_i\|^2 - \sum_{l=1}^{q} (x_k - v_i) d_{il}^T \right) \tag{9.37}$$

For $q = 1$, this is also called the *fuzzy c-lines (FCL)* clustering model. For FCV/FCL, U is then computed as

$$u_{ik} = 1 \left/ \sum_{j=1}^{c} \left(\frac{\|x_k - v_i\|^2 - \sum_{l=1}^{q}(x_k - v_i)d_{il}^T}{\|x_k - v_j\|^2 - \sum_{l=1}^{q}(x_k - v_j)d_{jl}^T} \right)^{\frac{1}{m-1}} \right. \tag{9.38}$$

V by (9.28), and D are the eigenvalues of the within-cluster covariance matrices (9.33).

$$d_{ij} = \text{eig}_j \sum_{k=1}^{n} u_{ik}^m (x_k - v_i)^T (x_k - v_i), \quad j = 1, \ldots, q \tag{9.39}$$

A third example of non-spherical prototypes (besides GK and lines/planes/hyperplanes) are *elliptotypes* [5] defined by the distance

$$d(x_k, v_i, d_{i1}, \ldots, d_{iq}) = \sqrt{\|x_k - v_i\|^2 - \alpha \cdot \sum_{l=1}^{q}(x_k - v_i)d_{il}^T} \tag{9.40}$$

with the parameter $\alpha \in [0, 1]$. For $\alpha = 0$ elliptotypes are points, and for $\alpha = 1$ elliptotypes are lines, planes, or hyperplanes. Elliptotypes are used, for example, to define the *fuzzy c-elliptotypes (FCE)* clustering model. Figure 9.9 (left) shows the cluster prototypes found by AO of FCE with the parameters $c = 2, m = 2, q = 1, \alpha = 0.5, t_{max} = 100$. The cluster centers v_1, v_2 are shown as circles, and the directions d_{11}, d_{21} are shown as arrows. They match well the local linear structures.

An example of more complicated geometric prototypes are circles. Each circle can be represented by a center v_i and a radius r_i. The distance between point x_k and such a circle is

$$d(x_k, v_i, r_i) = |\|x_k - v_i\| - r_i|. \tag{9.41}$$

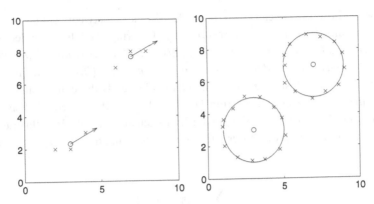

Fig. 9.9 Detection of lines and circles by clustering

which defines, for example, the *fuzzy c-shells (FCS)* clustering model [8]. In FCS, the centers are computed by (9.28), and the radii by

$$r_i = \frac{\sum\limits_{k=1}^{n} u_{ik}^m \|x_k - v_i\|}{\sum\limits_{k=1}^{n} u_{ik}^m} \tag{9.42}$$

Figure 9.9 (right) shows the FCS results for a data set with two circles. A wide variety of clustering models for complex geometries can be obtained by combining basic clustering models (CM, FCM, PCM, NC) with appropriate distance measures.

9.5 Relational Clustering

This section presents clustering methods that find clusters in relational data specified by symmetric distance matrices D, where $d_{ii} = 0$, $d_{ij} = d_{ji}$, $i, j = 1, \ldots, n$ [26]. The sequential clustering methods SAHN and SDHN from Sect. 9.2 consider distances between pairs of points are therefore suitable for relational clustering. The prototype-based clustering methods CM, FCM, PCM, and NC from the previous two sections need feature data $X \subset \mathbb{R}^p$ and are therefore not directly suitable for relational clustering.

Relational data may be mapped to feature data using multidimensional scaling or Sammon mapping (Chap. 4), and then prototype-based clustering methods may be used to find clusters in the resulting feature data. Here, instead, we consider modifications of prototype-based clustering algorithms that can be immediately applied to relational data, without explicitly producing feature data.

Prototype-based clustering models use distances between feature vectors and cluster centers. If we require the cluster centers to be chosen from the set of data vectors, $V \subseteq X$, the so-called *medoids* [22], then these clustering models can also be used for relational clustering, because all possible distances between feature vectors and medoids are available in D. The corresponding clustering models are called *relational c-medoids (RCMdd), relational fuzzy c-medoids (RFCMdd), relational possibilistic c-medoids (RPCMdd), relational noise clustering with medoids (RNCMdd)*. The restriction to medoids leaves the clustering model unchanged but adds the constraint $V \subseteq X$ that makes sure that U can be computed from D. The selection of the medoids V is a discrete optimization problem that is usually solved by exhaustive search. We illustrate this for RFCMdd. The contribution of cluster i to the objective function $J_{\text{RFCMdd}} = J_{\text{FCM}}$ is the additive term

$$J_i^* = \sum_{k=1}^{n} u_{ik}^m \| x_k - v_i \|^2 \tag{9.43}$$

so we can write $J = \sum_{i=1}^{c} J_i^*$. Without loss of generality we assume that $v_i = x_j$, so we have $\| v_i - x_k \| = d_{jk}$ and further

$$J_i^* = J_{ij} = \sum_{k=1}^{n} u_{ik}^m d_{jk}^2 \tag{9.44}$$

So the best choice of the medoids is $v_i = x_{w_i}$, $i = 1, \ldots, n$, where

$$w_i = \mathrm{argmin}\{J_{i1}, \ldots, J_{in}\} \tag{9.45}$$

This exhaustive search usually leads to a large computational effort.

A different approach to transform feature based clustering models to relational clustering models is to only implicitly compute the cluster prototypes, for example by inserting the equation for optimal cluster centers into the objective function of the clustering model. This so-called *reformulation* [15] yields objective functions for relational clustering. Reformulation of CM, FCM, PCM, and NC yields *relational c-means (RCM)*, *relational fuzzy c-means (RFCM)*, *relational possibilistic c-means (RPCM)*, and *relational noise clustering (RNC)* [16] . We illustrate the reformulation for the example of RFCM. For Euclidean distances inserting (9.28) into (9.25) yields the objective function

$$J_{\text{RFCM}}(U; D) = \sum_{i=1}^{c} \frac{\sum_{j=1}^{n} \sum_{k=1}^{n} u_{ij}^m u_{ik}^m d_{jk}^2}{\sum_{j=1}^{n} u_{ij}^m} \tag{9.46}$$

The necessary conditions for extrema of J_{RFCM} yield

$$
u_{ik} = 1 \left/ \sum_{j=1}^{n} \frac{\displaystyle\sum_{s=1}^{n} \frac{u_{is}^m d_{sk}}{\sum_{r=1}^{n} u_{ir}^m} - \sum_{s=1}^{n}\sum_{t=1}^{n} \frac{u_{is}^m u_{it}^m d_{st}}{2\left(\sum_{r=1}^{n} u_{ir}^m\right)^2}}{\displaystyle\sum_{s=1}^{n} \frac{u_{js}^m d_{sk}}{\sum_{r=1}^{n} u_{jr}^m} - \sum_{s=1}^{n}\sum_{t=1}^{n} \frac{u_{js}^m u_{jt}^m d_{st}}{2\left(\sum_{r=1}^{n} u_{jr}^m\right)^2}} \right.
\tag{9.47}
$$

$i = 1, \ldots, c, k = 1, \ldots, n$.

Imagine that the relational data matrix D may be computed from a feature data set X using the Euclidean norm. We will then call D a *Euclidean distance matrix*. Minimizing $J_{\mathrm{FCM}}(U, V; X)$ will then yield the same partition as minimizing $J_{\mathrm{RFCM}}(U; D)$. The same holds for the pairs CM/RCM, PCM/RPCM, NC/RNC, CMdd/RCMdd, FCMdd/RFCMdd, PCMdd/RPCMdd, and NCMdd/RNCMdd. It does not hold, if D is computed from X using a different norm or without using feature data at all. In this case, the minimization of the relational objective function might yield memberships $u_{ik} < 0$ or $u_{ik} > 1$. This can be avoided by transforming the non-Euclidean matrix D into a Euclidean matrix using the so-called *beta spread transformation* [14] that adds offsets $\beta \geq 0$ to all off-diagonal elements of D.

$$
D_\beta = D + \beta \cdot \cdot \begin{pmatrix} 0 & 1 & \ldots & 1 & 1 \\ 1 & 0 & \ldots & 1 & 1 \\ \vdots & \vdots & \ddots & \vdots & \vdots \\ 1 & 1 & \ldots & 0 & 1 \\ 1 & 1 & \ldots & 1 & 0 \end{pmatrix}
\tag{9.48}
$$

The parameter β may be initially set to zero and successively increased until all memberships are in the unit interval. For the corresponding clustering models with beta spread transformation we use the prefix "NE" for "non-Euclidean", for example *non-Euclidean relational fuzzy c-means (NERFCM)* [14] or *non-Euclidean relational possibilistic c-means (NERPCM)* [25].

The previously presented relational clustering models are derived from feature data clustering models considering hyperspherical data. In the previous section we have presented feature data clustering models for more complicated cluster shapes like ellipsoidals, hyperplanes, or circles. It is difficult to extend these models to relational data without generating feature data. For the support vector machine (Chap. 8) we have used a different approach to extend an algorithm from simple geometries (linear class borders) to more complicated geometries (nonlinear class borders): the *kernel trick*. For clustering the kernel trick is used to extend clustering algorithms for simple geometries (hyperspherical clusters) to more complicated geometries. The kernel trick has been applied, for example, to CM [12, 29, 38], FCM [34–36], PCM [37], NERFCM [17], and NERPCM [25]. We denote the kernelized variants with the prefix "k", for example kCM, kFCM, kPCM, kNERFCM, kNERPCM. In relational clustering with Euclidean distances the kernel trick yields

$$d_{jk}^2 = \|\varphi(x_j) - \varphi(x_k)\|^2 \tag{9.49}$$

$$= \big(\varphi(x_j) - \varphi(x_k)\big)\big(\varphi(x_j) - \varphi(x_k)\big)^T \tag{9.50}$$

$$= \varphi(x_j)\varphi(x_j)^T - 2\,\varphi(x_j)\varphi(x_k)^T + \varphi(x_k)\varphi(x_k)^T \tag{9.51}$$

$$= k(x_j, x_j) - 2 \cdot k(x_j, x_k) + k(x_k, x_k) \tag{9.52}$$

$$= 2 - 2 \cdot k(x_j, x_k) \tag{9.53}$$

where we assume that $k(x, x) = 1$ for all $x \in \mathbb{R}^p$. So a kernel variant of a non-Euclidean relational clustering model can be realized by a simple preprocessing step where the relational data matrix D is transformed to a *kernelized* matrix D' using (9.53), and the unchanged clustering algorithm is then applied to D' instead of D. For the Gaussian kernel (8.43) and the hyperbolic tangent kernel (8.44), for example, we obtain the kernel transformations

$$d'_{jk} = \sqrt{2 - 2 \cdot e^{-\frac{d_{jk}^2}{\sigma^2}}} \tag{9.54}$$

$$d'_{jk} = \sqrt{2 \cdot \tanh\left(\frac{d_{jk}^2}{\sigma^2}\right)} \tag{9.55}$$

Figure 9.10 shows these functions for $\sigma \in \{1, 2, 3, 4\}$. Notice the similarity between these two kernel transformations. Small distances are approximately linearly scaled, and the distances are softly clipped at the threshold $d' = \sqrt{2}$. All relation clustering methods presented here are invariant to an arbitrary scaling of the distances by a constant factor

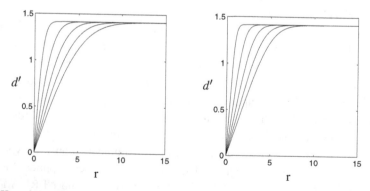

Fig. 9.10 Kernel transformation by Gaussian kernel (left) and hyperbolic tangent kernel (right)

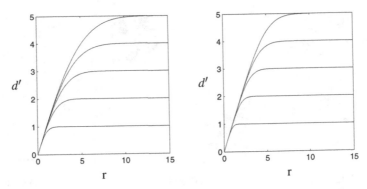

Fig. 9.11 Scaled kernel transformation by Gaussian kernel (left) and hyperbolic tangent kernel (right)

$\alpha > 0$, so D and D^*, $d_{jk}^* = \alpha \cdot d_{jk}$, $j, k = 1, \ldots, n$, will yield the same results. We can therefore multiply each function in Fig. 9.10 with an arbitrary constant factor, more specifically with a factor that yields a slope of one at the origin (Fig. 9.11). This view illustrates that the kernel transformations have the effect of a soft clipping of the distance at the threshold σ, so the effect of large distances in D, for example caused by inliers or outliers, is reduced [24].

9.6   Cluster Tendency Assessment

Clustering algorithms will find clusters in data, even if the data do not contain any clusters. *Cluster tendency assessment* attempts to find out if the data possess a cluster structure. A popular cluster tendency assessment method uses the *Hopkins index* [18, 19]. To compute the Hopkins index of a data set $X = \{x_1, \ldots, x_n\} \subset \mathbb{R}^p$ we first randomly pick m points $R = \{r_1, \ldots, r_m\}$ in the convex hull or (for simplicity) the minimal hypercube of X, where $m \ll n$. Then we randomly pick m data points $S = \{s_1, \ldots, s_m\}$ from X, so $S \subset X$. For both sets R and S we compute the distances to the nearest neighbor in X, which yields $d_{r_1}, \ldots, d_{r_m}$ and $d_{s_1}, \ldots, d_{s_m}$. Based on these distances the Hopkins index is defined as

$$h = \frac{\sum\limits_{i=1}^{m} d_{r_i}^p}{\sum\limits_{i=1}^{m} d_{r_i}^p + \sum\limits_{i=1}^{m} d_{s_i}^p} \tag{9.56}$$

which yields $h \in [0, 1]$. We distinguish three cases

1. For $h \approx 0.5$ the nearest neighbor distances in X are almost the same as between R and X, so R and X have similar distributions. Since R was purely randomly chosen, we can conclude that X is also randomly distributed and does not contain clusters.

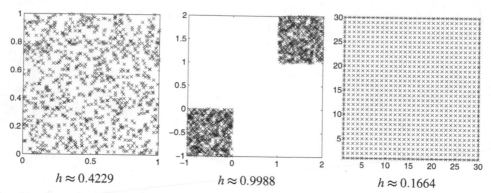

Fig. 9.12 Three data sets and their Hopkins indices

2. For $h \approx 1$ the nearest neighbor distances in X are relatively small, which indicates that X contains a cluster structure.
3. For $h \approx 0$ the nearest neighbor distances in X are relatively large, which indicates that X is on a regular grid.

To illustrate these three cases, Fig. 9.12 shows three different data sets and their Hopkins indices ($m = 10$, $n = 1000$). The left data set is randomly distributed, so $h \approx 0.5$. The middle data set contains two well-separated clusters, so $h \approx 1$. The right data set is on a regular rectangular grid, so h is very small.

9.7   Cluster Validity

All clustering methods presented here require to specify the number $c \in \{2, \ldots, n-1\}$ of clusters to find. In many applications the number of clusters is not known in advance. To estimate the number of clusters contained in the data, *cluster validity measures* are used. Clustering is performed for different numbers of clusters, and the resulting partition with the best validity indicates the appropriate number of clusters. Frequently used cluster validity measures are the average square membership or *partition coefficient (PC)* [2]

$$PC(U) = \frac{1}{n} \sum_{k=1}^{n} \sum_{i=1}^{c} u_{ik}^2 \tag{9.57}$$

and the average entropy or *classification entropy (CE)* [33]

$$CE(U) = -\frac{1}{n} \sum_{k=1}^{n} \sum_{i=1}^{c} u_{ik} \cdot \log u_{ik} \tag{9.58}$$

Table 9.1 Validity indices for crisp and indifferent partitions

U	$PC(U)$	$CE(U)$
$\begin{pmatrix} 1 & 0 & \dots & 0 \\ 0 & 1 & & 0 \\ \vdots & & \ddots & \vdots \\ 0 & 0 & \dots & 1 \end{pmatrix}$	1	0
$\begin{pmatrix} \frac{1}{c} & \dots & \frac{1}{c} \\ \vdots & \ddots & \vdots \\ \frac{1}{c} & \dots & \frac{1}{c} \end{pmatrix}$	$\frac{1}{c}$	$\log c$

where we define

$$0 \cdot \log 0 = \lim_{u \to 0} u \log u = 0 \tag{9.59}$$

Table 9.1 shows the validity indices PC and CE for a crisp (best) and for an indifferent (worst) partition. PC is max-optimal at one, so good choices of the number of clusters (or other algorithmic parameters) will yield PC values close to one; and CE is min-optimal at zero, so good choices will yield CE values close to zero. So, a good choice of the cluster number will yield a high value of PC (close to one), or a low value of CE (close to zero). Other validity measures (that may also be applied to nonfuzzy clusters) include the Dunn index, the Davis-Bouldin index, the Calinski-Harabasz index, the gap index and the silhouette index.

9.8 Self-organizing Map

A *self-organizing map* [20, 21, 32] is a heuristic method for mapping and clustering. The map is a regular q-dimensional array of a total of l reference nodes. Two-dimensional self-organizing maps may have a rectangular or hexagonal structure (Fig. 9.13). Each node has

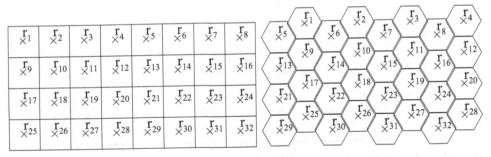

Fig. 9.13 Rectangular and hexagonal self-organizing map

a location vector $r_i \in \mathbb{R}^q$ that defines its position on the map, and a reference vector $m_i \in \mathbb{R}^p$ that relates to the data X, $i = 1, \dots, l$. For training the reference vectors $m_i \in \mathbb{R}^p$ are first randomly initialized. Then for each data point $x_k \in X$ we find the closest reference vector, say m_c, and update the reference vectors that are located on the map in the neighborhood of r_c. The neighborhood may be defined by the *bubble function*

$$h_{ci} = \begin{cases} \alpha(t) & \text{if } \|r_c - r_i\| < \rho(t) \\ 0 & \text{otherwise} \end{cases} \tag{9.60}$$

or the *Gaussian function*

$$h_{ci} = \alpha(t) \cdot e^{-\frac{\|r_c - r_i\|^2}{2 \cdot \rho^2(t)}} \tag{9.61}$$

The neighborhood radius $\rho(t)$ and the learning rate $\alpha(t)$ may be monotonically decreased, for example by

$$\alpha(t) = \frac{A}{B + t}, \quad A, B > 0 \tag{9.62}$$

Figure 9.14 shows the training algorithm of a self-organizing map. Notice the similarity of the update of the reference vectors with learning vector quantization (Chap. 8). After training the reference vectors or their nearest neighbor distances are visualized. Regions of similar (or even equal) reference vectors on the map indicate clusters in the data. Moreover, each data point can be mapped to the closest reference vector.

---

1. input: data $X = \{x_1, \dots, x_n\} \subset \mathbb{R}^p$,
   map dimension $q \in \{1, \dots, p-1\}$,
   node positions $R = \{r_1, \dots, r_l\} \subset \mathbb{R}^q$
2. initialize $M = \{m_1, \dots, m_l\} \subset \mathbb{R}^p$, $t = 1$
3. for each x_k, $k = 1, \dots, n$,

   a. find winner node m_c with

   $$\|x_k - m_c\| \le \|x_k - m_i\| \quad \forall i = 1, \dots, l$$

   b. update winner and neighbors

   $$m_i = m_i + h_{ci} \cdot (x_k - m_c) \quad \forall i = 1, \dots, l$$

4. $t = t + 1$
5. repeat from (3.) until termination criterion holds
6. output: reference vectors $M = \{m_1, \dots, m_l\} \subset \mathbb{R}^p$

Fig. 9.14 Training algorithm for a self-organizing map

Problems

9.1 Consider the data set $X = \{-6, -5, 0, 4, 7\}$.

(a) Draw the single linkage dendrogram.
(b) Draw the complete linkage dendrogram.
(c) Compute the sequence of cluster centers that c-means produces with initialization $V = \{5, 6\}$.
(d) Find an initialization for which c-means yields a different result for X.

9.2 Consider the relational data set

$$D = \begin{pmatrix} 0 & 1 & 4 \\ 1 & 0 & 2 \\ 4 & 2 & 0 \end{pmatrix}.$$

(a) Give the results of the c-medoids algorithm, $c = 2$, for all possible initializations of V.
(b) Why is relational c-means not suitable for this data set?
(c) How could the data be transformed to be used by relational c-means?
(d) Compute the transformed data set.
(e) Find a feature data set that corresponds to this transformed relational data set.

References

1. G. B. Ball and D. J. Hall. Isodata, an iterative method of multivariate analysis and pattern classification. In *IFIPS Congress*, 1965.
2. J. C. Bezdek. *Pattern Recognition with Fuzzy Objective Function Algorithms*. Plenum Press, New York, 1981.
3. J. C. Bezdek. Fuzzy models — what are they, and why? *IEEE Transactions on Fuzzy Systems*, 1(1):1–6, 1993.
4. J. C. Bezdek, C. Coray, R. Gunderson, and J. Watson. Detection and characterization of cluster substructure, I. Linear structure: Fuzzy c–lines. *SIAM Journal on Applied Mathematics*, 40(2):339–357, April 1981.
5. J. C. Bezdek, C. Coray, R. Gunderson, and J. Watson. Detection and characterization of cluster substructure, II. Fuzzy c–varieties and convex combinations thereof. *SIAM Journal on Applied Mathematics*, 40(2):358–372, April 1981.
6. J. C. Bezdek and R. J. Hathaway. Optimization of fuzzy clustering criteria using genetic algorithms. In *IEEE Conference on Evolutionary Computation, Orlando*, volume 2, pages 589–594, June 1994.
7. J. C. Bezdek, J. M. Keller, R. Krishnapuram, and N. R. Pal. *Fuzzy Models and Algorithms for Pattern Recognition and Image Processing*. Kluwer, Norwell, 1999.
8. R. N. Davé. Fuzzy shell clustering and application to circle detection in digital images. *International Journal on General Systems*, 16:343–355, 1990.

9. R. N. Davé. Characterization and detection of noise in clustering. *Pattern Recognition Letters*, 12:657–664, 1991.

10. W. H. E. Day and H. Edelsbrunner. Efficient algorithms for agglomerative hierarchical clustering methods. *Journal of Classification*, 1(1):7–24, 1984.

11. H. Fang and Y. Saad. Farthest centroids divisive clustering. In *International Conference on Machine Learning and Applications*, pages 232–238, 2008.

12. M. Girolami. Mercer kernel–based clustering in feature space. *IEEE Transactions on Neural Networks*, 13:780–784, 2002.

13. E. E. Gustafson and W. C. Kessel. Fuzzy clustering with a covariance matrix. In *IEEE International Conference on Decision and Control, San Diego*, pages 761–766, 1979.

14. R. J. Hathaway and J. C. Bezdek. NERF c–means: Non–Euclidean relational fuzzy clustering. *Pattern Recognition*, 27:429–437, 1994.

15. R. J. Hathaway and J. C. Bezdek. Optimization of clustering criteria by reformulation. *IEEE Transactions on Fuzzy Systems*, 3(2):241–245, May 1995.

16. R. J. Hathaway, J. W. Davenport, and J. C. Bezdek. Relational duals of the c–means algorithms. *Pattern Recognition*, 22:205–212, 1989.

17. R. J. Hathaway, J. M. Huband, and J. C. Bezdek. Kernelized non–Euclidean relational fuzzy c–means algorithm. In *IEEE International Conference on Fuzzy Systems*, pages 414–419, Reno, May 2005.

18. A. K. Jain and R. C. Dubes. *Algorithms for Clustering Data*. Prentice Hall, Englewood Cliffs, 1988.

19. A. K. Jain, M. N. Murty, and P. J. Flynn. Data clustering: A review. *ACM Computing Surveys*, 31(3):264–323, 1999.

20. T. Kohonen. Automatic formation of topological maps of patterns in a self–organizing system. In E. Oja and O. Simula, editors, *Scandinavian Conference on Image Analysis*, pages 214–220, Helsinki, 1981.

21. T. Kohonen. *Self–Organizing Maps*. Springer, Berlin, 2001.

22. R. Krishnapuram, A. Joshi, O. Nasraoui, and L. Yi. Low–complexity fuzzy relational clustering algorithms for web mining. *IEEE Transactions on Fuzzy Systems*, 9(4):595–607, August 2001.

23. R. Krishnapuram and J. M. Keller. A possibilistic approach to clustering. *IEEE Transactions on Fuzzy Systems*, 1(2):98–110, May 1993.

24. T. A. Runkler. The effect of kernelization in relational fuzzy clustering. In *GMA/GI Workshop Fuzzy Systems and Computational Intelligence, Dortmund*, pages 48–61, November 2006.

25. T. A. Runkler. Kernelized non–Euclidean relational possibilistic c–means clustering. In *IEEE Three Rivers Workshop on Soft Computing in Industrial Applications*, Passau, August 2007.

26. T. A. Runkler. Relational fuzzy clustering. In J. Valente de Oliveira and W. Pedrycz, editors, *Advances in Fuzzy Clustering and its Applications*, chapter 2, pages 31–52. Wiley, 2007.

27. T. A. Runkler. Wasp swarm optimization of the c-means clustering model. *International Journal of Intelligent Systems*, 23(3):269–285, February 2008.

28. J. Sander, M. Ester, H.-P. Kriegel, and X. Xu. Density-based clustering in spatial databases: The algorithm GDBSCAN and its applications. *Data Mining and Knowledge Discovery*, 2(2):169–194, 1998.

29. B. Schölkopf, A.J. Smola, and K. R. Müller. Nonlinear component analysis as a kernel eigenvalue problem. *Neural Computation*, 10:1299–1319, 1998.

30. P. Sneath and R. Sokal. *Numerical Taxonomy*. Freeman, San Francisco, 1973.

31. J. H. Ward. Hierarchical grouping to optimize an objective function. *Journal of American Statistical Association*, 58(301):236–244, 1963.

32. D. J. Willshaw and C. von der Malsburg. How patterned neural connections can be set up by self-organization. *Proceedings of the Royal Society London*, B194:431–445, 1976.

33. M. P. Windham. Cluster validity for the fuzzy c–means clustering algorithm. *IEEE Transactions on Pattern Analysis and Machine Intelligence*, PAMI-4(4):357–363, July 1982.

34. Z.-D. Wu, W.-X. Xie, and J.-P. Yu. Fuzzy c–means clustering algorithm based on kernel method. In *International Conference on Computational Intelligence and Multimedia Applications*, pages 49–54, Xi'an, 2003.

35. D.-Q. Zhang and S.-C. Chen. Fuzzy clustering using kernel method. In *International Conference on Control and Automation*, pages 123–127, 2002.

36. D.-Q. Zhang and S.-C. Chen. Clustering incomplete data using kernel–based fuzzy c–means algorithm. *Neural Processing Letters*, 18:155–162, 2003.

37. D.-Q. Zhang and S.-C. Chen. Kernel–based fuzzy and possibilistic c–means clustering. In *International Conference on Artificial Neural Networks*, pages 122–125, Istanbul, 2003.

38. R. Zhang and A.I. Rudnicky. A large scale clustering scheme for kernel k–means. In *International Conference on Pattern Recognition*, pages 289–292, Quebec, 2002.

Brief Review of Some Optimization Methods

In this book we discuss many data analysis methods that use some form of optimization. Many data analysis algorithms are essentially optimization algorithms. In this appendix we briefly review the basics of the optimization methods used in this book: optimization with derivatives, gradient descent, and Lagrange optimization. For a more comprehensive overview of optimization methods the reader is referred to the literature, for example [1–3].

A.1 Optimization with Derivatives

A popular approach to find extrema or saddle points of a differentiable function $y = f(x)$ is to set the necessary conditions for extrema of f to zero

$$\frac{\partial f}{\partial x^{(i)}} = 0 \qquad (A.1)$$

where the second derivative

$$\frac{\partial^2 f}{\partial (x^i)^2} \qquad (A.2)$$

is negative for maxima, positive for minima, and zero for saddle points. Suppose for example that we want to find the extrema of the function

$$y = f(x^{(1)}, x^{(2)}) = (x^{(1)} - 1)^2 + (x^{(2)} - 2)^2 \qquad (A.3)$$

© Springer Fachmedien Wiesbaden GmbH, part of Springer Nature 2020
T. A. Runkler, *Data Analytics*, https://doi.org/10.1007/978-3-658-29779-4

The derivatives of f are

$$\frac{\partial f}{\partial x^{(1)}} = 2(x^{(1)} - 1), \quad \frac{\partial f}{\partial x^{(2)}} = 2(x^{(2)} - 2) \tag{A.4}$$

$$\frac{\partial^2 f}{\partial (x^{(1)})^2} = 2, \quad \frac{\partial^2 f}{\partial (x^{(2)})^2} = 2 \tag{A.5}$$

Setting the derivatives of f to zero yields

$$2(x^{(1)} - 1) = 0 \quad \Rightarrow \quad x^{(1)} = 1, \quad 2(x^{(2)} - 2) = 0 \quad \Rightarrow \quad x^{(2)} = 2 \tag{A.6}$$

and since the second order derivatives are positive, f has a minimum at $x = (1, 2)$.

A.2   Gradient Descent

Gradient descent is an iterative approximative optimization method for differentiable functions. It is often suitable if setting the derivatives of f to zero cannot be explicitly solved. Gradient descent minimizes a function $y = f(x)$ by randomly initializing the parameter vector $x = x_0$ and then iteratively updating

$$x_k^{(i)} = x_{k-1}^{(i)} - \alpha \left. \frac{\partial f}{\partial x^{(i)}} \right|_{x=x_{k-1}} \tag{A.7}$$

with a suitable step length $\alpha > 0$ and $k = 1, 2, \ldots$. For our example (A.3) we initialize $x = (0, 0)$, use the gradients (A.4), and obtain with $\alpha = 1/2$

$$x_1^{(1)} = 0 - \frac{1}{2} \cdot 2 \cdot (0 - 1) = 1, \quad x_1^{(2)} = 0 - \frac{1}{2} \cdot 2 \cdot (0 - 2) = 2 \tag{A.8}$$

$$x_2^{(1)} = 1 - \frac{1}{2} \cdot 2 \cdot (1 - 1) = 1, \quad x_2^{(2)} = 2 - \frac{1}{2} \cdot 2 \cdot (2 - 2) = 2 \tag{A.9}$$

So in this case the gradient descent finds the correct minimum $x = (1, 2)$ in one step. For the same example $\alpha = 1/4$ yields

$$x_1^{(1)} = 0 - \frac{1}{4} \cdot 2 \cdot (0 - 1) = \frac{1}{2}, \quad x_1^{(2)} = 0 - \frac{1}{4} \cdot 2 \cdot (0 - 2) = 1 \tag{A.10}$$

$$x_2^{(1)} = \frac{1}{2} - \frac{1}{4} \cdot 2 \cdot (\frac{1}{2} - 1) = \frac{3}{4}, \quad x_2^{(2)} = 1 - \frac{1}{4} \cdot 2 \cdot (1 - 2) = \frac{3}{2} \tag{A.11}$$

$$x_3^{(1)} = \frac{3}{4} - \frac{1}{4} \cdot 2 \cdot (\frac{3}{4} - 1) = \frac{7}{8}, \quad x_3^{(2)} = \frac{3}{2} - \frac{1}{4} \cdot 2 \cdot (\frac{3}{2} - 2) = \frac{7}{4} \tag{A.12}$$

So the algorithm approximates of the correct minimum with a decreasing error that approaches zero as the number of steps approaches infinity. As a third example we choose $\alpha = 1$, which yields

$$x_1^{(1)} = 0 - 2 \cdot (0 - 1) = 2, \quad x_1^{(2)} = 0 - 2 \cdot (0 - 2) = 4 \tag{A.13}$$

$$x_2^{(1)} = 2 - 2 \cdot (2 - 1) = 0, \quad x_2^{(2)} = 4 - 2 \cdot (4 - 2) = 0 \tag{A.14}$$

So the algorithm infinitely alternates between $(2, 4)$ and $(0, 0)$. Finally, we choose $\alpha = 2$, which yields

$$x_1^{(1)} = 0 - 2 \cdot 2 \cdot (0 - 1) = 4, \quad x_1^{(2)} = 0 - 2 \cdot 2 \cdot (0 - 2) = 8 \tag{A.15}$$

$$x_2^{(1)} = 4 - 2 \cdot 2 \cdot (4 - 1) = -8, \quad x_2^{(2)} = 8 - 2 \cdot 2 \cdot (8 - 2) = -16 \tag{A.16}$$

$$x_3^{(1)} = -8 - 2 \cdot 2 \cdot (-8 - 1) = 28, \quad x_3^{(2)} = -16 - 2 \cdot 2 \cdot (-16 - 2) = 56 \tag{A.17}$$

So the algorithm diverges to $(\pm\infty, \pm\infty)$. Thus, gradient descent is a simple optimization method that works well if the step length α is chosen appropriately. Choosing α can be avoided by considering the second derivatives, as in *Newton's method* that finds extrema of quadratic functions in one single step.

$$x_k^{(i)} = x_{k-1}^{(i)} - \frac{\left.\frac{\partial f}{\partial x^{(i)}}\right|_{x=x_{k-1}}}{\left.\frac{\partial^2 f}{\partial (x^{(i)})^2}\right|_{x=x_{k-1}}} \tag{A.18}$$

A method to minimize linear functions under constraints is the *simplex* or *Nelder-Mead algorithm*. *Sequential programming* is a family of optimization methods that successively minimizes *approximations* of the function to be minimized, for example sequential *linear* programming or sequential *quadratic* programming.

A.3   Lagrange Optimization

Consider the problem of optimizing a function $y = f(x)$ under the constraint $g(x) = 0$. For example, we may require that $x^{(1)} = x^{(2)}$ by defining the constraint function $g(x^{(1)}, x^{(2)}) = x^{(1)} - x^{(2)}$. In some cases constrained optimization can be done by inserting the constraint into the function to be optimized and then performing (unconstrained) optimization. For example, if we consider again our example (A.3) with the constraint $x^{(1)} = x^{(2)}$, then we can replace $x^{(2)}$ by $x^{(1)}$ and obtain

$$y = f(x^{(1)}) = (x^{(1)} - 1)^2 + (x^{(1)} - 2)^2 = 2(x^{(1)})^2 - 6x^{(1)} + 5 \qquad \text{(A.19)}$$

with the first and second order derivatives

$$\frac{\partial f}{\partial x^{(1)}} = 4x^{(1)} - 6, \quad \frac{\partial^2 f}{\partial (x^{(1)})^2} = 4 \qquad \text{(A.20)}$$

Setting the (first order) derivative of f to zero yields

$$4x^{(1)} - 6 = 0 \quad \Rightarrow \quad x^{(1)} = \frac{3}{2} \qquad \text{(A.21)}$$

so under the constraint $x^{(1)} = x^{(2)}$, f has a minimum at $x = (\frac{3}{2}, \frac{3}{2})$. In many cases inserting the constraint does not lead to an explicitly solvable function. In such cases *Lagrange optimization* may be used. Lagrange optimization transforms the problem of optimizing a function $y = f(x)$ under the constraint $g(x) = 0$ into the unconstrained problem of optimizing a function

$$L(x, \lambda) = f(x) - \lambda g(x) \qquad \text{(A.22)}$$

with an additional parameter $\lambda \in \mathbb{R}$. For our example (A.3) with $g(x^{(1)}, x^{(2)}) = x^{(1)} - x^{(2)}$ we obtain

$$L(x^{(1)}, x^{(2)}, \lambda) = (x^{(1)} - 1)^2 + (x^{(2)} - 2)^2 - \lambda(x^{(1)} - x^{(2)}) \qquad \text{(A.23)}$$

with the derivatives

$$\frac{\partial L}{\partial x^{(1)}} = 2(x^{(1)} - 1) - \lambda, \quad \frac{\partial L}{\partial x^{(2)}} = 2(x^{(2)} - 2) + \lambda, \quad \frac{\partial L}{\partial \lambda} = -(x^{(1)} - x^{(2)}) \qquad \text{(A.24)}$$

Setting the derivatives to zero yields

$$x^{(1)} = 1 + \frac{\lambda}{2}, \quad x^{(2)} = 2 - \frac{\lambda}{2}, \quad x^{(1)} = x^{(2)} \qquad \text{(A.25)}$$

and further $x = (\frac{3}{2}, \frac{3}{2})$. For multiple constraints $g_1(x), \ldots, g_r(x)$, Lagrange optimization is defined as

$$L(x, \lambda) = f(x) - \sum_{i=1}^{r} \lambda_i g_i(x) \qquad \text{(A.26)}$$

with the parameters $\lambda_1, \ldots, \lambda_r \in \mathbb{R}$. Optimization becomes considerably more difficult when inequality constraints are part of the problem, for example, $g(x) \geq 0$. In this case we use the Karush-Kuhn-Tucker (KKT) theory. See [2] for this method.

References

1. R. Fletcher. *Practical Methods of Optimization*. Wiley, Chichester, 2000.
2. J. Nocedal and S. J. Wright. *Numerical Optimization*. Springer, New York, 2006.
3. R. K. Sundaram. *A First Course in Optimization Theory*. Oxford University Press, Oxford, 1996.

Solutions

Problems of Chapter 2

2.1

(a) interval, ratio
(b) interval, ratio
(c) interval, ratio
(d) ordinal, interval, ratio
(e) nominal, ordinal, interval, ratio
(f) interval, ratio

2.2

(a) circle with radius 1
(b) rhombus with the edge points $(0, \pm 1)$ and $(\pm 1, 0)$
(c) coordinate axes without origin
(d) square with the edge points $(\pm 1, \pm 1)$ and $(\pm 1, \mp 1)$
(e) horizontal ellipse through the points $(0, \pm 1)$ and $(\pm\sqrt{10}, 0)$
(f) $\begin{pmatrix} x & y \end{pmatrix} \cdot \begin{pmatrix} 1 & -2 \\ 0 & 1 \end{pmatrix} \cdot \begin{pmatrix} x \\ y \end{pmatrix} = (x - y)^2 = 1 \Rightarrow y = x \pm 1$

2.3

(a) for example $\{\oplus \oplus \oplus, \ominus \ominus \ominus\}$: Hamming distance 3, edit distance 3
(b) for example $\{\oplus \ominus \oplus, \ominus \oplus \ominus\}$: Hamming distance 3, edit distance 2 (remove first, append last)
(c) all pairs whose Hamming distance is different from their edit distance: $\{\oplus\ominus\oplus, \ominus\oplus\ominus\}$, $\{\oplus\oplus\ominus, \oplus\ominus\ominus\}$, $\{\ominus\ominus\oplus, \ominus\oplus\oplus\}$; so the solution are all sets of sequences that contain not more than one sequence of any of these three sets, and that may or may not contain

© Springer Fachmedien Wiesbaden GmbH, part of Springer Nature 2020
T. A. Runkler, *Data Analytics*, https://doi.org/10.1007/978-3-658-29779-4

the remaining sequences $\ominus\ominus\ominus$ and $\oplus\oplus\oplus$, for example $\{\ominus\ominus\ominus, \oplus\ominus\oplus, \oplus\oplus\ominus, \ominus\ominus$ $\oplus, \oplus\oplus\oplus\}$

2.4

(a) $\dfrac{x+y}{\sqrt{2(x^2+y^2)}} = 1 \Rightarrow x^2+2xy+y^2 = 2x^2+2y^2 \Rightarrow x^2-2xy+y^2 = 0 \Rightarrow$
$(x-y)^2 = 0 \Rightarrow x = y$ except $x = y = 0$

(b) $\dfrac{x+y}{\sqrt{2(x^2+y^2)}} = \dfrac{1}{\sqrt{2}} \Rightarrow x^2+2xy+y^2 = x^2+y^2 \Rightarrow 2xy = 0 \Rightarrow x =$
0 or $y = 0$ except $x = y = 0$

(c) $\dfrac{x+y}{\sqrt{2(x^2+y^2)}} = \sqrt{\dfrac{9}{10}} \Rightarrow x^2+2xy+y^2 = \frac{9}{5}x^2+\frac{9}{5}y^2 \Rightarrow \frac{4}{5}x^2-2xy+\frac{4}{5}y^2 =$
$0 \Rightarrow x^2 - \frac{5}{2}xy + y^2 = 0 \Rightarrow x = \frac{5}{4}y \pm \sqrt{\frac{25}{16}y^2 - y^2} = \frac{5}{4}y \pm \frac{3}{4}y =$
$2y$ or $\frac{1}{2}y$ except $x = y = 0$

(d) (positive parts of) two lines through the origin with angles α from the x and y axes, where α increases with decreasing similarity: $0°$ (minimum) for similarity 1, $22.5°$ (mean) for similarity $0.3\sqrt{10}$, $45°$ (maximum) for similarity $\frac{1}{2}\sqrt{3}$

Problems of Chapter 3

3.1

(a) stochastic errors: noise, deterministic errors: outlier 1.03
(b) stochastic errors: measurement noise, deterministic errors: wrong format $1.03 \leftrightarrow 1030$
(c) $(920, 950, 950)$
(d) noise is reduced, outlier is removed

3.2

(a) FIR
(b) none
(c) IIR

3.3

(a) $y = (0, 0, 1, 2, 2, 2, 2, 2)$
(b) $y = (0, 0, 1, 2+a, 2+2a+b, 2+2a+2b, 2+2a+2b, 2+2a+2b)$
(c) never
(d) $2+2a+2b = 0 \Rightarrow a+b = -1$

Problems of Chapter 4

4.1

(a) eigenvectors $(\frac{\sqrt{2}}{2}, -\frac{\sqrt{2}}{2}), (\frac{\sqrt{2}}{2}, \frac{\sqrt{2}}{2})$

(b) $Y = \{-\frac{3}{2}\sqrt{2}, -\frac{3}{2}\sqrt{2}, 0, \frac{3}{2}\sqrt{2}, \frac{3}{2}\sqrt{2}\}$

(c) $e = \frac{1}{5}(\frac{1}{2} + \frac{1}{2} + 0 + \frac{1}{2} + \frac{1}{2}) = \frac{2}{5}$

(d) $D^X = \begin{pmatrix} 0 & \sqrt{2} & \sqrt{5} & 3\sqrt{2} & 2\sqrt{5} \\ \sqrt{2} & 0 & \sqrt{5} & 2\sqrt{5} & 3\sqrt{2} \\ \sqrt{5} & \sqrt{5} & 0 & \sqrt{5} & \sqrt{5} \\ 3\sqrt{2} & 2\sqrt{5} & \sqrt{5} & 0 & \sqrt{2} \\ 2\sqrt{5} & 3\sqrt{2} & \sqrt{5} & \sqrt{2} & 0 \end{pmatrix}$, $D^y = \begin{pmatrix} 0 & 0 & \frac{3}{2}\sqrt{2} & 3\sqrt{2} & 3\sqrt{2} \\ 0 & 0 & \frac{3}{2}\sqrt{2} & 3\sqrt{2} & 3\sqrt{2} \\ \frac{3}{2}\sqrt{2} & \frac{3}{2}\sqrt{2} & 0 & \frac{3}{2}\sqrt{2} & \frac{3}{2}\sqrt{2} \\ 3\sqrt{2} & 3\sqrt{2} & \frac{3}{2}\sqrt{2} & 0 & 0 \\ 3\sqrt{2} & 3\sqrt{2} & \frac{3}{2}\sqrt{2} & 0 & 0 \end{pmatrix}$

$\Rightarrow$ four points at $(\sqrt{2}, 0)$ (twofold), $(\sqrt{5}, \frac{3}{2}\sqrt{2})$ (fourfold), $(3\sqrt{2}, 3\sqrt{2})$ (twofold), $(2\sqrt{5}, 3\sqrt{2})$ (twofold)

(e) multiplication with E^T

(f) $y = e^{-\|x - (\frac{\sqrt{2}}{2}, -\frac{\sqrt{2}}{2})\|^2}$

(g) $Y = \{e^{-6-3\sqrt{2}}, e^{-6-3\sqrt{2}}, e^{-\frac{1}{2}}, e^{-6+3\sqrt{2}}, e^{-6+3\sqrt{2}}\}$

(h) because $\|y_3 - y_1\| = \|y_3 - y_2\| \neq \|y_3 - y_4\| = \|y_3 - y_5\|$

4.2

(a) $q \geq 3$

(b) maximum 6 points on the positive main diagonal

(c) $d_{34}^x > d_{31}^x + d_{14}^x$, so D^X is not Euclidean

4.3

(a) $x' = (atanh\, y, atanh\, y)$

(b) $Y = \{tanh\, 0, tanh\, \frac{1}{2}, tanh\, \frac{1}{2}, tanh\, 1\}$, $X' = \{(0, 0), (\frac{1}{2}, \frac{1}{2}), (\frac{1}{2}, \frac{1}{2}), (1, 1)\}$,

$e = \frac{1}{4}\left(0 + (\sqrt{2}/2)^2 + (\sqrt{2}/2)^2 + 0\right) = \frac{1}{4}$

(c) one-dimensional PCA

Problems of Chapter 5

5.1

(a) spurious correlation: both features depend on temperature; use partial correlation without influence of temperature

(b) coincidence; nothing

(c) nonlinear dependency; use chi-square test

5.2

(a) scatter plot $\Rightarrow$ linear slope $= 0 \Rightarrow$ correlation $= 0$

(b) $H = \begin{pmatrix} 1\ 1 \\ 1\ 1 \\ 1\ 1 \end{pmatrix} \Rightarrow$ low correlation

(c) $H = \begin{pmatrix} 0\ 4\ 0 \\ 1\ 0\ 1 \end{pmatrix} \Rightarrow$ high correlation

5.3 $\dfrac{h_{ijk}}{n} = \dfrac{h_i^{(1)}}{n} \cdot \dfrac{h_j^{(2)}}{n} \cdot \dfrac{h_k^{(3)}}{n} \quad \Rightarrow \quad h_{ijk} = \dfrac{h_i^{(1)} \cdot h_j^{(2)} \cdot h_k^{(3)}}{n^2}$

$\Rightarrow \quad \chi^2 = \dfrac{1}{n} \sum\limits_{i=1}^{r} \sum\limits_{j=1}^{s} \sum\limits_{k=1}^{t} \dfrac{\left(n^2 \cdot h_{ijk} - h_i^{(1)} \cdot h_j^{(2)} \cdot h_k^{(3)}\right)^2}{h_i^{(1)} \cdot h_j^{(2)} \cdot h_k^{(3)}}$

Problems of Chapter 6

6.1

(a) symmetry $\Rightarrow f(x) = 0$

(b) symmetry $\Rightarrow f(x) = \dfrac{-1+1+15+1-1}{5} = 3$

(c) $\varepsilon = 4 \Rightarrow$ only y_5 in the linear case $\Rightarrow 2 \cdot 4 \cdot (y_5' - 3) - 16 = (y_5 - 3)^2 = 12^2 \Rightarrow$ $y_5' = 23$

(d) in robust regression the same effect is obtained by a much larger outlier, so robust regression is more robust to outliers

6.2

(a)

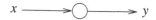

$$x \longrightarrow \bigcirc \longrightarrow y$$

(b)

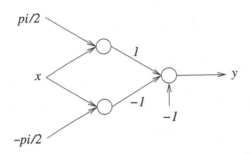

(c)

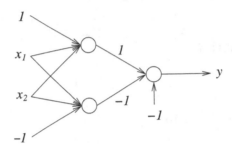

6.3

(a) $h_1 = x_1 + x_2, h_2 = x_1 - x_2, h_3 = 0$
  $y_1 = ah_1 + bh_2 + ch_3 = (a + b)x_1 + (a - b)x_2$
  $y_2 = dh_1 + eh_2 + fh_3 = (d + e)x_1 + (d - e)x_2$
(b) $a = b = d = \frac{1}{2}, e = -\frac{1}{2}$ or $a = d = e = \frac{1}{2}, b = -\frac{1}{2}$
(c) advantage: x_1 and x_2 are preserved; disadvantage: x_3 gets lost

6.4 overfitting

Problems of Chapter 7

7.1

(a) $\mu = 10000, \sigma = \sqrt{\frac{1}{2}(5000^2 + 0 + 5000^2)} = 5000, x = (-1, 0, 1)$
(b) $x_t = ax_{t-1} + b$
  $x_2 = ax_1 + b \Rightarrow 0 = -a + b$
  $x_3 = ax_2 + b \Rightarrow 1 = b \Rightarrow a = 1$
  $\Rightarrow x_t = x_{t-1} + 1$

(c) $x = (-1, 0, 1, 2, 3)$, $y = (5000, 10000, 15000, 20000, 25000)$
(d) $y_\infty \to \infty$

7.2

(a) $x_t = tanh(x_{t-1} + 1)$
$x_0 = -1$
$x_1 = tanh(-1 + 1) = 0$
$x_2 = tanh(0 + 1) \approx \frac{3}{4}$
$x_3 = tanh(\frac{3}{4} + 1) \approx \frac{15}{16}$
$y = (5000, 10000, 13750, 14687.50)$
(b) $x_\infty = tanh(x_\infty + 1)$
$1 = tanh(1 + 1) \Rightarrow x_\infty = 1 \Rightarrow y_\infty = 15000$

Problems of Chapter 8

8.1

(a) $p(1 \mid (0, 0)) = \frac{2}{3}$, $p(2 \mid (0, 0)) = \frac{1}{3}$

(b) $x \cdot \begin{pmatrix} 1 \\ 1 \end{pmatrix} - \frac{1}{2} = 0$

(c) $f(x) = \begin{cases} 1 & \text{if } x^{(1)} < 0.5 \text{ and } x^{(2)} < 0.5 \\ 2 & \text{if } x^{(1)} > 0.5 \text{ or } x^{(2)} > 0.5 \\ \text{undefined} & \text{otherwise} \end{cases}$

(d) $f(x) = 2$
(e) naive Bayes, SVM, NN: TPR=1, FPR=0; 3-NN: TPR=1, FPR=1

8.2

(a) limit: $p(\oplus) \cdot p(x \mid \oplus) = p(\ominus) \cdot p(x \mid \ominus)$ $\Rightarrow$ $\frac{1}{2\pi} \cdot \frac{1}{1+(x+1)^2} = \frac{1}{2\pi} \cdot \frac{1}{1+(x-1)^2}$ $\Rightarrow$
$1 + (x + 1)^2 = 1 + (x - 1)^2$ $\Rightarrow$ $x^2 + 2x + 2 = x^2 - 2x + 2$ $\Rightarrow$ $x = 0$
maximum of $p(x \mid \oplus)$ at $x = -1 \Rightarrow$ "positive" for all $x < 0$.
(b) $(1 - p)(x^2 + 2x + 2) = p(x^2 - 2x + 2)$ $\Rightarrow$ $(2p - 1)x^2 - 2x + 4p - 2 = 0$ $\Rightarrow$
$x^2 - \frac{2}{2p-1}x + 2 = 0$ $\Rightarrow$ $x = \frac{1}{2p-1} \pm \sqrt{\frac{1}{(2p-1)^2} - 2}$
limit when $\sqrt{\cdots} = 0$ $\Rightarrow$ $\frac{1}{2p-1} = \pm\sqrt{2}$ $\Rightarrow$ $p = \frac{1}{2} \pm \frac{\sqrt{2}}{4} \approx 15\%, 85\%$
always "positive" for $p > 85\%$

(c) $p < 15\%$: never "positive"

$p > 85\%$: always "positive"

otherwise: "positive" for $x \in [\frac{1}{2p-1} - \sqrt{\frac{1}{(2p-1)^2} - 2}, \frac{1}{2p-1} + \sqrt{\frac{1}{(2p-1)^2} - 2}]$

8.3

(a) no

(b) $\varphi(x)\varphi(y)^T = k(x, y) = (xy^T)^2 = (x_1 y_1 + x_2 y_2)^2 = x_1^2 y_1^2 + 2x_1 x_2 y_1 y_2 + x_2^2 y_2^2 =$
$(x_1^2, x_2^2, x_3)(y_1^2, y_2^2, y_3)^T = x_1^2 y_1^2 + x_2^2 y_2^2 + x_3 y_3 \quad \Rightarrow \quad x_3 = \sqrt{2} x_1 x_2, y_3 =$
$\sqrt{2} y_1 y_2 \quad \Rightarrow \quad \varphi(x) = (x_1^2, x_2^2, \sqrt{2} x_1 x_2)$

(c) mappings $Y_- = \{(1, 1, \sqrt{2}), (1, 1, \sqrt{2})\}$, $Y_+ = \{(1, 1, -\sqrt{2}), (1, 1, -\sqrt{2})\}$
normal vector $w = (0, 0, 1)$, offset $b = 0$, margin $2\sqrt{2}$

8.4

(a) $H = -\frac{1}{4} \log_2 \frac{1}{4} - \frac{1}{2} \log_2 \frac{1}{2} - \frac{1}{4} \log_2 \frac{1}{4} = 1.5$ bit

(b) $H(90 \text{ kW}) = -\frac{1}{2} \log_2 \frac{1}{2} - \frac{1}{2} \log_2 \frac{1}{2} = 1$ bit

$H(120 \text{ kW}) = 0$ bit

expected value $\frac{1}{2} \cdot 1$ bit $+ \frac{1}{2} \cdot 0$ bit $= 0.5$ bit, gain 1.5 bit $- 0.5$ bit $= 1$ bit

(c) $H(\text{gas}) = -\frac{1}{3} \log_2 \frac{1}{3} - \frac{1}{3} \log_2 \frac{2}{3} \approx \frac{19}{36} + \frac{14}{36} = \frac{33}{36} = \frac{11}{12}$

$H(\text{diesel}) = 0$ bit

expected value $\frac{3}{4} \cdot \frac{11}{12}$ bit $+ \frac{1}{4} \cdot 0$ bit $= \frac{11}{16}$ bit, gain $\frac{24}{16}$ bit $- \frac{11}{16}$ bit $= \frac{13}{16}$ bit < 1 bit

(d) first ask for kW; if 90 kW, then ask for gas or diesel

Problems of Chapter 9

9.1

(a) $\{\{-6, -5\}, \{0, \{4, 7\}\}\}$

(b) $\{\{\{-6, -5\}, 0\}, \{4, 7\}\}$

(c) $V = \{-\frac{7}{4}, 7\}$, $V = \{-\frac{11}{3}, \frac{11}{2}\}$

d) for example $V = \{-\frac{11}{2}, \frac{11}{3}\}$

9.2

(a) $V = \{x_1, x_2\} \Rightarrow U = \begin{pmatrix} 1 & 0 & 0 \\ 0 & 1 & 1 \end{pmatrix}$,

$\quad V = \{x_1, x_3\} \Rightarrow U = \begin{pmatrix} 1 & 1 & 0 \\ 0 & 0 & 1 \end{pmatrix}$,

$\quad V = \{x_2, x_3\} \Rightarrow U = \begin{pmatrix} 1 & 1 & 0 \\ 0 & 0 & 1 \end{pmatrix}$

(b) not Euclidean: $d_{12} + d_{23} < d_{13}$

(c) β spread transformation

(d) $1 + \beta + 2 + \beta = 4 + \beta \quad \Rightarrow \quad \beta = 1$

$\quad \Rightarrow \quad D' = \begin{pmatrix} 0 & 2 & 5 \\ 2 & 0 & 3 \\ 5 & 3 & 0 \end{pmatrix}$

(e) for example $X' = \{0, 2, 5\}$

Index

© Springer Fachmedien Wiesbaden GmbH, part of Springer Nature 2020
T. A. Runkler, *Data Analytics*, https://doi.org/10.1007/978-3-658-29779-4

Printed in the United States
By Bookmasters